£1·50

Watches in Colour

Watches in Colour

RICHARD GOOD

BLANDFORD PRESS
POOLE *DORSET*

First published in 1978
by Blandford Press Ltd,
Link House, West Street, Poole,
Dorset BH15 1LL

ISBN 0 7137 0831 X

Colour printed by Sackville Press, Billericay
Text printed in Great Britain by
Fletcher & Son Ltd, Norwich and bound by
Richard Clay (The Chaucer Press) Ltd,
Bungay, Suffolk

Contents

1 The Beginnings of the Watch

Watches developed from small clocks. The exact moment at which they can be described as a watch must, by the very nature of things, be uncertain.

By definition a watch must be able to be worn or carried on the person. What shape or how small must the timepiece be before it is reasonable to carry it about? Who can say?

When documentary evidence is all that one has to go on then this must be accepted as a guide. With this in mind, it is probable that watchmaking began in Italy. In November 1462, Bartolomeo Manfredi wrote to the Marquess Lodovico Gonzana Modena that he had begun to assemble a small watch similar to that owned by the Duke of Modena, and that he was determined to perfect it. The term that he employed is specific because in Italian 'orologetto' does not mean a small clock but definitely a watch. This point is agreed by both ancient and modern dictionaries.

The clocks that undoubtedly led to the watch were the tambour or drum clocks. Here the dial was horizontal and had no protective cover. Although weighing half a pound such clocks began to be carried, either suspended from the neck or placed in a leather pouch, which nobles and gentlemen wore at the waist. They were unreliable as sooner or later the hand would catch, stopping the clock. These small clocks had lips at the top and bottom for aesthetic reasons and when the idea arose of adding a cover to protect the hand, the upper part of the case band was left smooth and the cover edge appeared as the lip. Unfortunately, no early Italian watches have survived, a great misfortune for the horological historian.

One of the earliest paintings in which a watch appears is by Hans Eworth. It is a portrait dated 1563 of Lord Darnley, later to be the husband of Mary, Queen of Scots. He is shown wearing what appears to be a gold and blue enamel circular watch on a cord around his neck. We learn from her will that the Queen of Scots left to Lord Darnley – 'One watch garnished with ten diamonds, two rubies and a cord of gold.'

Early watches were undoubtedly costly items. They were extremely difficult to make and must have taxed the ability of the finest available craftsmen. Every tooth of each wheel had to be marked out and filed by hand. The craftsman himself usually had to make the files that he used. Mainsprings were particularly difficult to make and soon became the province of specialists.

Wherever watchmaking began, it is almost certain that Nuremberg was the great centre of watchmaking in the early years of the 16th century. Later in France in about 1525, other centres came into being notably at Dijon, Rouen, Blois and Paris. The watches made in Germany and France were very similar, the main difference being in the fitting of the striking arrangement. German watches had the 'nags head' method of striking, whereas the French system had 'warning'. All striking of this period was of the count wheel variety.

Religious strife led to the exodus of French Huguenot watchmakers. They settled in various places, one being Geneva where they helped to found the centre that was eventually to become so well known. In the last two decades of the 16th century more French craftsmen left France for London and stimulated the watchmaking industry there. Only a small number of craftsmen were needed to form a centre of watch production. They needed to take little with them but their skill and could set up anywhere, providing that the necessary supply of metals and the know-how to work them already existed.

Watches – The Train and Escapement

All the wheelwork in mechanical watches is powered by a coiled spring, usually contained in a barrel. The inner end of the spring is attached to an arbor and the outer end attached to the wall of the barrel. When the arbor or the barrel is turned, a ratchet prevents the spring from immediately unwinding again. The only way it can unwind and thus expend the energy that has been put into it, is to turn the train of gears to which its output end is connected. This may be the barrel or the arbor according to the design.

In an ordinary watch the first of the gears to be turned is the centre wheel, which is mounted on a pinion. This centre wheel usually carries the minute hand. The last of the gears in the train is the escapement wheel. By reason of the high ratio between the barrel and the escape wheel, only a small portion of the power of the mainspring reaches it. If the rate of rotation of this escape wheel can be accurately controlled, then so will the speed of rotation of the rest of the gear train. Thus, the hand will rotate at the correct speed and indicate the right time. Further gearing can be employed to achieve any other rate of rotation required, once in twelve hours, once a minute, once a year or what you will.

The mechanism that controls the rate of rotation of the escape wheel is the escapement. This comprises all the elements that follow and includes the escape wheel. The most important is the balance; the foliot will be discussed later.

The balance wheel is pivoted on an axis and is constructed so that the major portion of its weight lies in its rim. This is to make maximum use of its inertial properties. In most of the early watches, where there was a balance, this had no spring, but after about 1650 balance springs were fitted, which gave the balance a specific rate of oscillation. One end of the spring was attached to the piece that supported and located the balance and the other to the balance axis, or staff, as it is known. If the balance is turned through a certain angle it will immediately be urged back to its rest position by the spring. However, the inertia of the balance will cause it to pass the rest position and, but for inevitable frictional losses, it would oscillate for ever. As the energy is given to the spring through coiling, it is transferred to energy contained in the balance because of its motion. It is the energy contained in the escape wheel that is used to make up the frictional losses that occur at the balance.

If the balance oscillations could be made free of factors that upset their uniformity, a perfect timekeeper would result. Unfortunately, this is not possible and the following are those factors that upset timekeeping:

1. The fact that the impulse cannot be given exactly at the rest position of the balance and spring assembly.

2. Changes in the balance losses that need to be made good due to changes in temperature, changes in oil viscosity and changes in position of the timekeeper.

3. Changes in the motive power itself.

4. Changes in temperature that affect the balance and spring itself by altering the moment of inertia of the balance and the strength of the balance spring.

Escapement Types

It is evident that some escapements must give better results than others. There are two basic types of escapements. The first, the frictional rest type and the second, the detached escapement.

The frictional rest type operates in such a way that once a tooth of the escape wheel has finished impulsing the balance the following tooth rests on some part of the balance or its axis. Evidently this cannot be a good type of escapement since the balance is never free from interference.

The detached escapement is one in which after a tooth of the escape wheel has finished giving impulse, the following tooth is arrested by some part, other than the balance, and kept away from the balance until the next time it is necessary for the balance to release the escape wheel. This way the inference to the balance can be kept as far as is possible around the desirable position, when most of the energy of the system is in the moving balance.

Endless thought and experiment has been put into the attainment of the perfect escapement, balance and spring. The answer is still being sought, but is likely to become purely academic due to the advent of the electronic watch.

The Mainspring

What made portable timepieces possible then, was the invention of the mainspring as a source of power. Previously the falling of a weight had supplied power to most mechanical clocks but obviously this was not suitable for a portable timekeeper.

It is probable that locksmiths pioneered the development of the mainspring, a ribbon of hardened and tempered steel, which provides power by releasing the energy that has been put into it to deform it. Such springs were very difficult to make at this early stage of the engineering art. Indeed, even today, they are still the province of specialists, for although some present-day craftsmen make every other part of a watch, not one would contemplate manufacturing his own mainsprings.

The mainspring, however, has one unfortunate characteristic that presents a great problem; that is, when fully wound it gives a greater force than when it is run down. This was a tremendous disadvantage in early watches, since their controlling element, the foliot or balance, depended on a constant source of power for its ability to keep any sort of time at all. To help somewhat, stop work was used to prevent the worst condition which occurred when the spring was fully wound. The stop work ensured that the watch could only be wound at the most to within about one turn of fully wound, although sometimes only two turns of the mainspring output were utilised, these being around the half-wound condition. It is here that the output of a spring is at its most uniform.

Two devices were developed to overcome the difficulty, one was called the stackfreed and the other the fusee. The stackfreed (Fig. 1) consisted of a stiff blade spring (A), which bore against a cam (B), that was geared to the barrel arbor (C). This spring resisted the force of the mainspring to a greater or lesser degree so as to help to equalise the spring output.

The earliest datable stackfreed watch is in a tambour case and is in the Louvre, Paris. It was made by Viet Schauffer of Munich in 1554. It is probable that the stackfreed was only introduced a few years before this date. Watches usually went for about twelve to fourteen hours so that they needed to be wound twice a day. Stop work remained and was an integral part of the stackfreed, the tooth that was uncut forming the stop. In later stackfreeds the cam against which the blade spring acted was made of brass and screwed to the stop wheel. It could thus be removed to alter its shape without altering the set-up of the stop work.

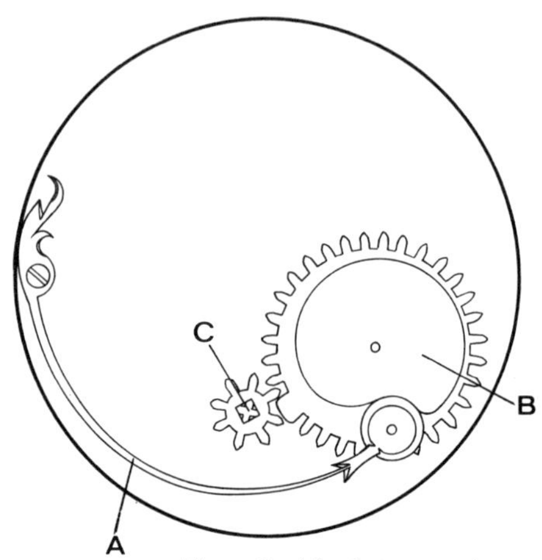

Fig. 1 Stackfreed.

A stackfreed watch with an English name does exist, although it is not certain that the signature is genuine. The name on the dial is G. Smith and there are also the initials GS on the movement. The stackfreed cam works for 270° of its rotation. It retards the mainspring for 20°, acts frictionally for a further 50° and then actively assists the mainspring for the remaining 200°. The retarding part of the cam covers 70° but some of this is prevented from acting due to the stop work. The cam is mounted on a stop wheel having twenty-six teeth which meshes with a pinion of eight leaves, thus allowing an effective three and a quarter turns of the mainspring. It is not possible to say that the style indicates anything other than the fact that it might have been made in Germany or the Netherlands. The watch also has hog's bristle regulation which is to be discussed later. All of the other extant stackfreeds are thought to have been made in Germany, Switzerland or the Netherlands. The stackfreed was a very crude device and did not last for long.

The fusee, however, which is in effect a pulley of varying diameter, was a sophisticated and elegant answer to the problem of the varying

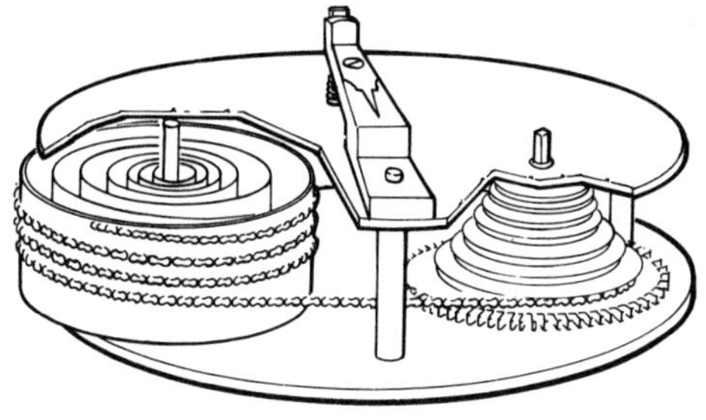

Fig. 2 Fusee.

power of the mainspring, so much so that it is still in use to this day. As can be seen from Fig. 2, the mainspring when fully wound works on the smallest diameter. In the figure, however, it is run down and is on the largest diameter of the continuous spiral groove machined on the fusee. As the mainspring runs down, then, it works on an increasing radius so as to offset its loss of power.

2 The Foliot and the Verge Escapement

As already described, the escapement in a watch or clock is that part that transforms the rotation of the train of gears into the reciprocating motion required by the controlling part of the mechanism. The first mechanical escapement was the verge, which was probably used to begin with as a type of alarm ringing device in monasteries although both its origin and its date of invention are obscure.

Two paddle-shaped pieces, known as flags, are mounted on a shaft that runs across the escape wheel, which has upright standing

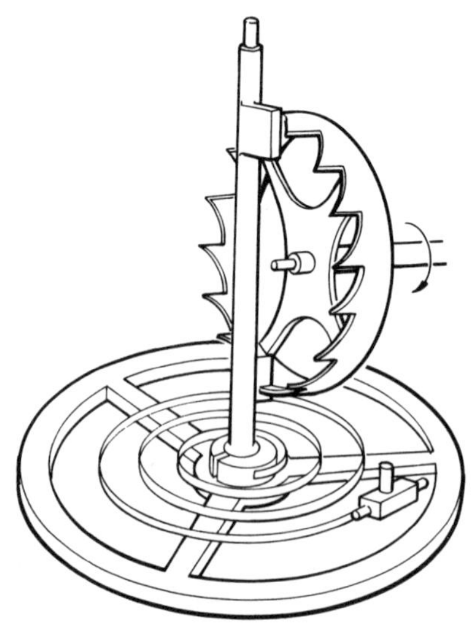

Fig. 3 Verge with crown wheel, balance and spring.

teeth. There are usually an odd number of teeth in the escape wheel. Due to the obvious resemblance, the escape wheel is known as a crown wheel (see Fig. 3). The flags mesh with the teeth so that one or the other is always in contact with a tooth. Thus, as the wheel turns, the flags, and thus the shaft, move first one way and then the other. The speed at which the wheel rotates depends upon the force applied to it and the inertial resistance of the shaft. This inertial resistance is increased by mounting a bar or a balance wheel on the shaft. In early clocks the bar, or foliot as it is called, often carried small weights that could be moved along it, so as to change the timekeeping, but this was not applicable to watches. Instead, the foliot ended in two fixed weights. The balance wheel has an advantage over the foliot in watches in as much as it is easier to 'poise'; that is, to make adjustments so that when it runs on edge no heavy point exists around its periphery to further upset timekeeping. Because of this only very early watches have a foliot, or bar balance.

The Hog's Bristle

Some control of the timekeeping of clocks with foliots was possible by moving the weights on the arm, so that they came nearer to or further away from its centre of rotation, thus altering the moment of inertia. This method of regulation was not practicable in watches.

In some early watches the balance arm was made to bank against a hog's bristle, as in the 'English' stackfreed watch already mentioned. By altering the position of the bristle and thus the angle at which the arm first touched it, a degree of adjustment is possible.

Watches were also regulated by altering the set-up of the mainspring, first by setting up the mainspring through a ratchet and ratchet wheel on the barrel arbor (see Plate 4). A later arrangement where the arbor that engages with the mainspring has a wheel mounted upon it, that can be turned by means of a worm, is shown in Plate 11c. The end of this worm was squared and if the watch went fast over a period, as it so often did, then it could be slowed by setting the mainspring up slightly, using the winding key. This system had dangers, for if the mainspring is set up too far, the fusee stop work could become inoperative and the gut line or chain broken.

Stop Work

Stop work is absolutely essential with a fusee, since the full force that is able to be applied by the winding key must never be directly felt by the gut line or chain. If this is allowed to happen then the line can be broken. The line can overcome the resistance of the mainspring, of course, because this is a design requirement and the amount of the resistance is known. Therefore, there must be a device to limit the amount of winding so that this ceases before the mainspring is

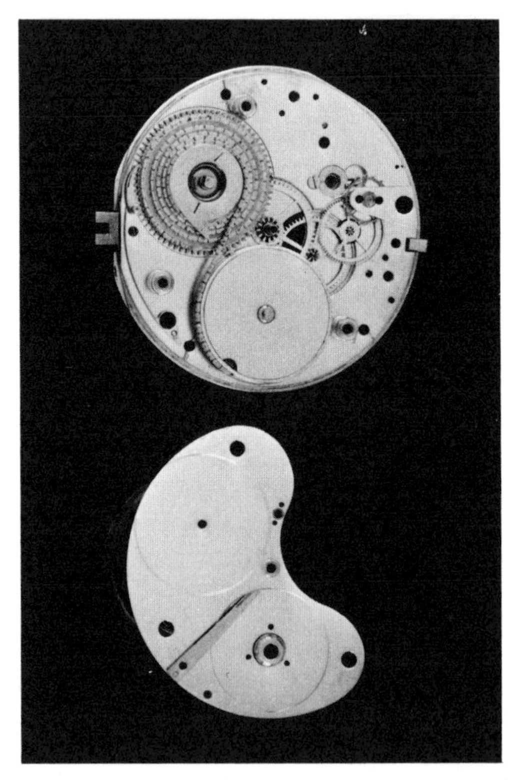

Fig. 4 Reverse fusee and stopwork. The beak of the piece on top of the fusee catches on the notched end of the spring on the top plate when this is pushed up by the chain. This happens as the last groove of the fusee is filled by the chain.

fully wound. This is accomplished by a hook that is pushed into the way of a projection on the fusee by the line, as it fills the last groove on the fusee. A version is shown in Fig. 4, and is associated with a reverse fusee.

Stop work is also applied to the going barrel, that is the ordinary barrel that drives the wheelwork directly, by means of a gear on its periphery. However, this is not required for the same reason, as the mainspring and its hooks are able to take the force applied by winding with a normal key when the mainspring is fully wound. Here stop work, if fitted, is used to eliminate the very high torque given by the mainspring when fully wound. Accordingly, further winding is prevented when there is still about half a turn to a turn (or even more) of winding to go, before the fully wound condition is reached. Stop work that accomplishes this is shown in Plates 44 and 47. It also prevents the mainspring from running down fully, although there is no great virtue in this.

At the middle of the 17th century the watch as a timekeeper was a poor affair, having errors of up to half an hour a day. Some radical improvement, a step forward, like that which occurred when the clock was transformed by the pendulum, was also required for the watch.

However, let us take pause here to examine the decorative aspects of the watch before, during and after this step – the invention of the balance spring – took place.

3 National Styles

The Movement

The difference in striking trains between early German and French watches has already been noted. When English watches appeared at the end of the 16th century they followed the French system of striking. German watches often had an alarum in preference to striking-work. Sometimes this alarum mechanism was separate and could be attached at will.

Early plate pillars were plain and square or round section. The top plate was pinned to secure it, a practice which continued well into the 19th century. Even very early watches had one screw. The stackfreed spring was usually secured by a screw and so was the set-up ratchet in a fusee watch (see Plate 9).

By the end of the 16th century national styles began to develop. German watches developed slowly. Iron frames and the stackfreed continued in use throughout the first quarter of the 17th century, when the Thirty Years War (1618–48) so crippled Germany that she ceased to be a factor in the horological field. This left France supreme, although her position was to be challenged by the British in the last quarter of the century.

By 1600 the movement was a beautiful item, although its time-keeping had not noticeably improved. Errors of a quarter of an hour a day were commonplace. Since timekeeping was so bad, attention was drawn away from it by lavish decoration, both of the case and the movement, and by adding complications such as calendar and astronomical work.

In French watches the balance cock was enlarged and was pierced and engraved. It was pinned to the plate. Pierced and engraved parts were also used to cover the locking plate of the striking train, alarum stop work and the set-up ratchet. The maker's name also began to be engraved on the plate.

In Britain, the balance cock was given floral-type decoration and for a little while after 1600 a border was engraved around the top-plate edge. After about 1620, the balance cock began to be secured by a screw instead of being pinned. As the century went on the shape changed from oval to circular and covered the balance completely. Pillars became more decorative. The Egyptian type was the commonest and after 1600 the tulip shape began to appear as did other types of decoration between the movement plates. A three-wheel train and twice-a-day winding remained the rule until about 1675. During the third quarter of the 17th century, thin watches appeared. The gut line also gave way to the fusee chain, although this is rarely found before 1670.

As to the escapement, there was no serious rival to the verge, so that, apart from the decorative aspects of watches, the period from the invention of the watch to the introduction of the balance spring was a period of stagnation. However, from the collector's standpoint and the external appearance of the watch, it is indeed a very interesting period.

Clock and Watchmakers' Guilds had existed abroad from early in the 16th century, but nothing much in this direction was achieved in Britain until 1631, when the Worshipful Company of Clockmakers was founded. David Ramsey was its first Master. From this time onward the quality of British watchmaking improved until, by 1675 at the time of the development of the balance spring, the British makers were able to take the lead from the French.

After the introduction of the balance spring it soon became difficult to make a weak enough mainspring to cope with a three-wheel train so another wheel was introduced, making a going period of twenty-six hours virtually universal. Some makers, both French and English – Tompion was one – thought they could dispense with the fusee, but soon found it necessary to return to using it with the verge escapement. The set-up regulator that the owner needed to manipulate was, however, found to be unnecessary but was retained for the convenience of the maker and repairer.

The application of the pendulum to clocks made a great impression on people and when the balance spring came into use this fact was still fresh in their minds. The new watches often sold better if there was a pendulum connotation. In France, watches were made with solid cocks and a simulated pendulum bob that could be seen swinging through an annular slot in this cock (see Plate 21). It was in fact no more than a disc on the balance arm. With English watches the balance was placed between the back plate and the dial, and this 'bob' appeared through a slot in the dial. This practice, however, did not last much after 1690.

Soon after the introduction of the balance spring, the decoration of the English watch movement became fairly standardised. The cock changed from the floreate pattern to an arabesque pattern, at first bold, but becoming increasingly fussy. The table covering the balance was circular and later was to have a solid rim. The foot was not so regular in shape, but by 1690 both foot and table had acquired a well-delineated rim, that of the foot following the plate edge. After about 1680 most clocks had a mask engraved where the table joins the foot; prior to 1685 this mask was small. This mask did not disappear until the middle of the 19th century where verge watches were concerned. Up to about 1740 the pattern of the decoration on the cock was symmetrical, but as the rococo style took over it became asymmetrical.

The French seldom used any pillar other than the Egyptian type in the *Oignons* and plain baluster pillars in their later thin watches, where the movement did not hinge out of the case. The cursive script of the English maker's name became less bold after 1680 and soon after 1690 appeared more and more often in plain capitals, as it almost always did in France. Dust caps over English movements are not often found before 1725 and rarely before 1715. After 1725 however, they became the rule for watches of any quality. Usually of gilt brass, they were, however, sometimes made of silver.

After 1725 the cock became more and more solid and engraved instead of pierced. When compensation balances were introduced, the table of the balance cock ceased to be circular and became wedge-shaped. Sometimes these cocks were solid and engraved, but pierced cocks (including Arnold's) continued up to the end of the century. Verge watches, however, continued to have circular tables until the mid-19th century.

The Case, Dial and Hands

Up to 1650, it is easier to date a watch by the style and decoration than it is by the mechanism. Plain metal cases are the only ones known before 1600. Gold and silver cases were made, records show this, but none survive; only gilt brass cases are extant.

Fig. 5 This German mid-16th century watch has a tambour case and movement with stackfreed and foliot type balance.

Since so many watches were clock-watches or alarums most cases had to be pierced to let out the sound. Also the lid had to be pierced to avoid the necessity of raising it to tell the time (see Fig. 5). These cases were usually cast. Most watches were shaped like boot polish tins. The bell was fixed to the bottom and the movement was hinged and swung out of the front for winding. Winding from the front, or through a hole in the back and bell, did not come until the second half of the 17th century. After casting, the case was usually chiselled in quite high relief. As the 17th century came to an end the chiselling became less deep, and engraving began to take its place. During the last quarter of the 17th century the cover began to be domed and the sides slightly convex, instead of straight. Some cases were octagonal but in general cases were circular (see Fig. 6).

Initially the dial was almost always engraved gilt metal. The applied silver chapter ring does not really appear until after 1600.

Twenty-four-hour dials were used in Italy and Germany. The first twelve hours were marked in Roman numerals, the second in Arabic. IIII was used for aesthetic reasons, not IV, and the Arabic 2 was Z (see Plate 1). After 1600 when English watches appeared these had the normal 2, but the German watches continued with the Z for another two decades or so. The centre of the dial usually had a star-shaped pattern.

Fig. 6 This watch has a dial with enamelled cartouches and a strong single hand.

The single hands were of steel or iron and strongly made to resist the forces put upon them (see Fig. 6) during hand setting, since they were pushed round by the finger. The only exception to the ferrous metal hand is found in enamelled watches where the hand is gilt metal. The pendant was usually fixed, drilled through from back to front and fitted with a loose ring (see Plate 1).

During the 16th century cases were made in a great variety of shapes, styles and decoration. In the first half of the 17th century the pair case appeared. The arrival of expensive and delicate cases made a second protective case, in which to keep the watch when not in use, a sensible idea. At first these were made of leather, but with the introduction of pockets it became usual to also wear the outer case. By the middle of the 17th century outer cases were an integral part of the watch and not meant to be separated from it.

The dial side of the outer case was open and the outer case itself soon became decorated. By 1670 the anomalous position was then

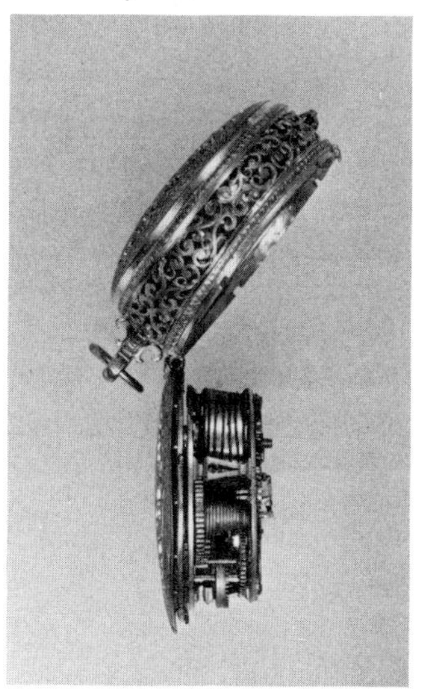

Fig. 7 This watch has an early gut fusee, Egyptian pillars, and pierced band to the case.

achieved when the inner case became completely plain and the outer sometimes so decorated that, if delicate, it was itself protected with a third case. From 1680 onwards the ordinary English watch had two plain cases, the outer to prevent the ingress of dirt through the winding hole. This resulted in a thick watch. The French, however, did not follow this system but had a single case, the watch being wound through the dial. This has not been a good thing for posterity since so many French watches have damaged dials as a result.

Around 1600 the form watch appeared, the case in the form of a skull, a book, a dog, crucifix, bird, flower bud, etc. (see Plates 2, 3, 7 and 18 and Figs. 8a and 8b). The traditional round, octagonal and increasingly oval watches continued to be made throughout the first quarter of the 17th century. Straight sides to the case increasingly gave way to the convex side, and in British watches particularly the whole case became like a somewhat flattened egg, a shape no doubt conducive to comfortable wear in the pocket. These cases were not

Fig. 8a The front of the watch shown in Plate 18.

Fig. 8b The movement of the watch shown in Plate 18.

decorated and were to become known as 'Puritan' watches by the middle of the century, by which time the dial also had become plain and the case almost invariably silver. Glass covers had also appeared during the second quarter of the 17th century. Gilt brass remained the usual metal, but silver and, to a lesser extent, gold cases also appeared.

French watches began to take on the elongated oval or the octagonal shape and this style was followed by the Swiss watches that began to appear. After the third quarter of the 17th century German watches virtually disappeared. True minute hands are seldom found before 1680.

During the third quarter of the 17th century the watch began to assume the form it was to retain for the next century. The matt surface silver or gold dial began to appear and the figures of the chapter ring grew longer. By 1675 they had become so long as to overwhelm the dial (see Plate 19). Hands became simpler, sometimes being quite plain; most had tails although many are found without. The pendant ring began to change from its original position, lying at right angles to the case, and moved to its present position lying in the same plane as the case. This probably followed the introduction of waistcoats into England by Charles II.

The balance spring brought about great changes in the English watch when it became generally introduced in 1675. The emphasis moved from exterior decoration to the aspects connected with timekeeping ability and the commonest case was a plain pair case. During the third quarter of the century watches had become much thinner but changed once the emphasis was on timekeeping. The French watch became especially thick and earned for itself the name of 'oignon' (onion). The commonest French case was a single gilt-brass case, decorated in a similar way to the case with the hour numerals on enamel cartouches (see Fig. 6). There was sometimes a narrow ring inside the hour plaque with the hours and half-hours marked on it. Minute hands were unusual and did not become universal until after 1700. When there is a minute hand there is sometimes another chapter ring outside the numerals with minute marks. Hands continued to be pushed about with the finger and even after 1700 may be found to have no motion work, so that they have to be set separately. Winding is through a hole in the dial, sometimes through the hand centre.

Back to English watches. When cases were of gold these were hallmarked and of twenty-two carat gold. Silver cases were seldom hallmarked before 1740. Watchmakers began to number the movements and the case makers to use their initials. Repoussé cases made their appearance, but before 1715 this decoration appears as radial fluting. Repoussé scenes and figures do not appear much until 1725. The great period for repoussé cases lasted from then until 1750, after which it again declined in favour until by 1770 such cases were rarely found (see Plate 27).

Christopher Pinchbeck discovered an alloy of three parts zinc and four of copper that bears his name. The secret of its composition was kept from the time of its discovery in 1720 until the end of the century. The industrial spy was not yet a breed to be reckoned with – or so it would appear.

Shagreen-covered cases came into fashion and the popularity of pinwork on leather, horn and shell continued (see Plate 31). Tortoise-shell was also used and was often inlaid with silver or gold strips. The shape of the decoration is cut out of the shell and the shell then heated and the strips are pressed in.

Pendants continued to be loose rings until about 1690 then the hinged stirrup type became almost universal. The case hinges also changed from the earlier square-ended type to the curved ends that merge more into the case rim.

Dials

Dials became standardised after 1700, after a very interesting period during which makers experimented with the best way of telling the minutes as well as the hours.

The concentric minute and hour hand had appeared by 1680 but was not universally accepted. These other methods were tried:

1. The wandering-hour dial (see Plate 23).
2. The differential dial.
3. The six-hour dial (see Plate 24).

They all try to tell the time by means of a single indicator. These types of dial are discussed where they appear with the appropriate watches.

A strange thing about the wandering-hour dials is that nearly all of them have a royal connection, such as a royal portrait on the dial, or the royal arms engraved on the cock, or both. Also many have the fluted repoussé case which is rare at this time. As one example bears a portrait of Queen Anne it is evident that this type of dial continued to be made well after the turn of the century. Seconds dials are occasionally found at a very early date in the normal modern position.

Hands

The hands are either 'beetle and poker' (see Fig. 9) or else the hour hand sometimes 'tulip'. The tulip hand did not survive long after 1715, but the beetle and poker went on until after 1800.

Dials were silver or gold with a matt ground and champlevé hour and minute numerals. The maker's name, when it appeared, was usually on a polished cartouche in the centre area of the dial. Enamel dials are not common before 1725 on English watches, when Graham adopted them as his standard dial.

For the next fifty years after Graham invented the cylinder

Fig. 9 This watch movement has an enamelled dial with beetle and poker hands. It is also centre seconds.

escapement in about 1725, another period of little change occurred. Both Continental and British watches hardly altered during this time.

Dials did change slowly. The old concentric circles gradually disappeared as did the minute marks, no longer needed as people began to read the time at a glance. By the end of the century the minute marks had virtually disappeared.

Engine turning began to appear from 1770 onward, probably because of the influence of Breguet. This was to be often associated with enamelling – the combination known as basse-taille enamelling.

A new school of precision makers came into being; Arnold and Earnshaw in England and Breguet in France. These makers stamped the watch of the time with surety. Elegant severity became the keynote in England. So much was timekeeping in mind that Emery sometimes used the clock-type dial that had previously been applied to regulators. Here, the minute circle occupied the whole dial with the minute hand on its own at the centre of the watch. Subsidiary dials at twelve and six o'clock were for hours and seconds. Emery used a special spade hand (see Fig. 10). This arrangement was much copied for the next forty years. Both Arnold and Emery, in common with other first-class makers, used the consular case in which the front and back meet in such a way that no band is visible, although in common watches, the pair case remained in vogue until the end of the century.

Soon after 1800 a plain matt gold dial with raised and polished numerals became fashionable in England, for watches with cylinder and duplex escapements. Serpentine hands often went with these dials. On the Continent, painted enamel dials of poor quality became common. Often they were allied to simple automata such as a revolving windmill (see Plate 22).

During the 1830–40s the period of heavy decoration began. Heavy cast and engraved watch cases with four-colour gold dials were impressive indeed and felt as good as they looked. Without the superb workmanship that went into them they could have seemed merely vulgar.

Automata movements were often repeaters, where the figures

Fig. 10 This watch shows the spade hands peculiar to Josiah Emery.

moved when the watch was made to strike (see Plates 35, 39, 41). Sometimes, these figures were engaged in pleasurable but far from innocent pastimes. Form watches came back into fashion in Switzerland soon after 1800 and have continued to be made ever since.

4 Breguet and his Influence

The man who really changed styles on the Continent was Breguet. He set up on his own in 1782 and at once established a new style of restrained elegance. His eye for proportion was unfailing but throughout his work he also had function firmly in mind.

One of his most elegant watches was the 'souscription', although it was his cheapest. This had a flat-sided band with plain rim bezels. Both back and glass were almost flat. The body and back were usually silver, the bezels and pendant ring gold. The back might be either plain or engine turned (see Plate 30).

His fashionable watches were almost always engine turned and were often quite thin. The case was not remarkable, having a narrow rounded band and curved bezels. Breguet did not make much use of enamelling, mainly reserving enamelled cases for his montres à tact (see Plate 46) and for watches for the Turkish market.

For his complicated watches, Breguet used the consular type of case where there was effectively no band. The sides of the case might be nearly flat or curved. Breguet mainly used red gold like most other Continental makers of this period. Inside the back cover there should be a 'B' with the watch and case number. There were French hallmarks at this time, but they are difficult to interpret and date.

Dials were of enamel, gold or silver. The enamel dials usually had Arabic numerals. Roman numerals on an enamel dial were rare in Breguet's lifetime. The signature appears below the 6 position in capital letters; either 'Breguet' or 'Breguet et Fils' after about 1807, but not invariably. The seconds dial is not a separate piece, for enamel dials were not in vogue in Breguet's lifetime. The seconds dial appears anywhere on a Breguet dial, its position was dictated by the movement design. This indifference to symmetry in the dial sets Breguet apart from most other makers. Breguet did not vary his hand design to any extent either so much so that his type of hand bears his name to this day. These hands can be seen in Plate 36; they are also called 'moon hands'. Metal dials are signed in the same way as the enamel dials. The numerals are Roman and the chapter ring plain. The dial centre is patterned with simple engine turning and, if present, the seconds dial is sunk. Soon after Breguet returned to Paris after the Revolution he discovered that his work was being copied and his signature forged. The result was his 'secret signature' intended to protect him and his customers. Put on both enamel and silver dials, although not invariably, the signature was engraved with a diamond point from a master plate using a pantograph. The name Breguet was usually followed by the watch number and is positioned just below the 12 above the fixing screw. On metal dials the signature appears twice, to the right and left and just below the 12.

After Breguet

From 1820 onwards little happened to the outward form of the watch. It is as if everything that happened within the watch prevented anything from happening to the outside. The only changes that took place were mainly connected with the change in winding methods.

There were one or two exceptions, the lady's watch of 1840 for instance. There were a few fashions, false Gothic, then false Greek, followed after the opening of the Suez Canal by false Egyptian styles. After about 1880 there was a return to the Renaissance style, somewhat modified by the revival of the Louis XV style.

In the late 1800s a new style was developed. This was Art Nouveau. The movement which did so much to popularise Art Nouveau was fostered by the enthusiasms and beliefs of the Arts and Crafts Society at whose head was the talented William Morris. The society was worried about what they felt to be the erosion of standards of craftsmanship by the increase of machine production methods. They achieved only a limited success and the only really important effect of the movement was the introduction into England of the Art Nouveau style from the Continent.

The greatest exponents of the style among the European jewellers were Georg Jensen in Copenhagen and Lalique in Paris. The Art Nouveau style was described by its followers as a return to naturalism. Chief among the motifs were flowers, the tulip and daffodil, but the female form and head were also often depicted. The representations were lively and sensual and characterised by a flowing grace that soon imprints itself on the eye as the signature of the styles.

For a while the movement and interest in it faded, but there has been for some years a greatly renewed interest that has resulted in the best examples of Art Nouveau being treated as if they were truly antique, that is conforming to the artificial requirement of being more than one hundred years old. Art Nouveau, although it affected clocks more, did have a corresponding effect on watches, as can be seen clearly by the illustrations in the book of the 1900 Paris Exhibition.

1 A gilt metal alarum watch pre-balance spring and verge escapement. Probably German, dating from the third quarter of the 16th century.

2 A silver gilt fritillary watch, *c.* 1620, with gut fusee, by Daniel Van Pilcom of Amsterdam.

3 A silver skull watch, *c.* 1620, by J. C. Vvolf, with verge escapement and pre-balance spring.

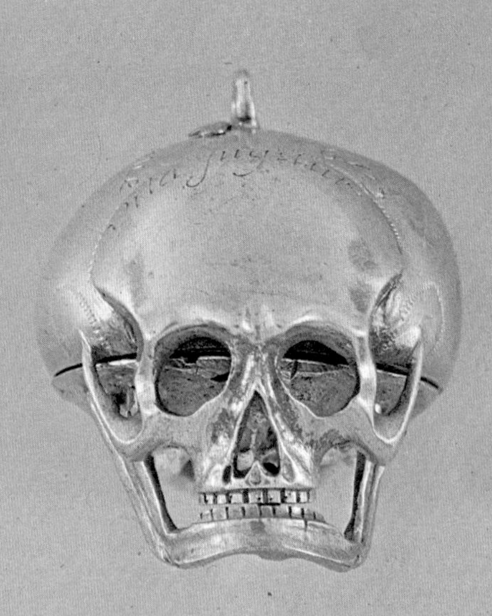

4 A Swiss astronomical watch, *c.* 1620, with verge escapement and pre-balance spring, by Gaspard Girod of Geneva.

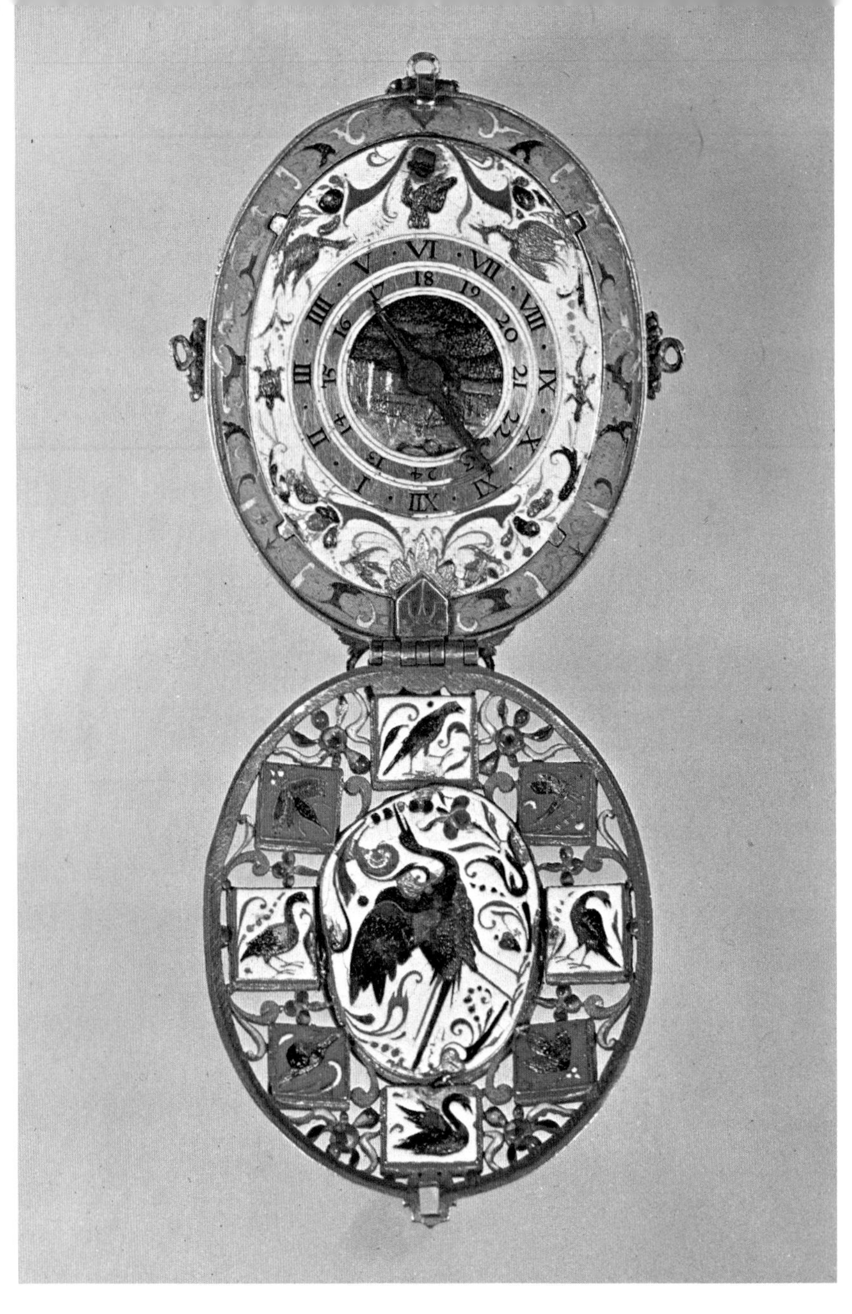

5 A French verge watch pre-balance spring, *c.* 1620. It is in an elaborately enamelled and jewelled case.

6 A Swiss verge watch pre-balance spring, by J. Sermand of Geneva (1595-1651). The crystal case has an enamelled rim and bezel.

7 A French crucifix verge watch pre-balance spring by O. Tinelly, 1630-35. The case is embossed silver and gilt.

8 This enamelled gold-cased verge watch pre-balance spring was made by D. Bouquet of London. The case is probably Swiss and the watch dates from 1630-40.

9 This tiny French verge watch with pre-balance spring was made by A. Bretonneau of Paris, 1638-43. The enamelled gold case measures only 22½mm by 20mm.

10 This German astronomical verge watch has a pendulum balance with a straight hairspring. It was made by George Seydell in the mid-17th century, and has a silver case.

11a An English verge watch pre-balance spring.

11b The gold and enamelled case for the watch shown opposite was probably made in Geneva.

11c Its movement was made by Isaac Pluvier of London, 1641-65. Note the pinned balance cock and worm set-up.

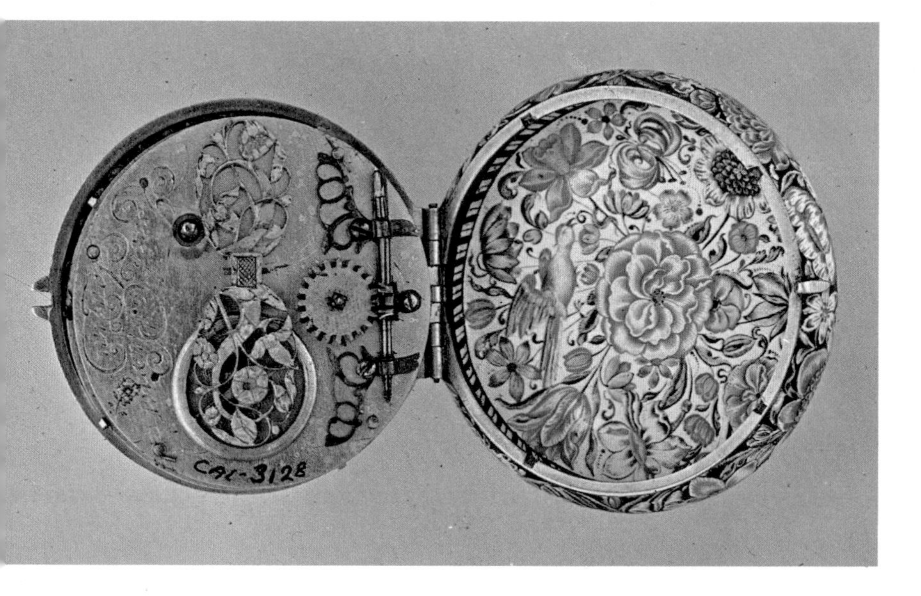

12 A Swiss verge watch, *c.* 1684, by Estienne Ester. The enamelled back shows Venus and Adonis.

13 This French verge watch was made by B. Foucher, 1630-40. The enamelling is attributed to Jean or Henri Toutin and shows scenes of the Amazons.

14 *Opposite.* Beautiful miniatures in enamel from the edge of a French watch case.

15 This Swiss clock watch with alarm was made by Jean Baptiste Duboule of Geneva (1615-94), who was also responsible for its engraving.

16 This pre-hairspring German verge watch was made for the Turkish market and thus has Arabic numerals. It dates from *c.* 1680 and has a gilt finish case.

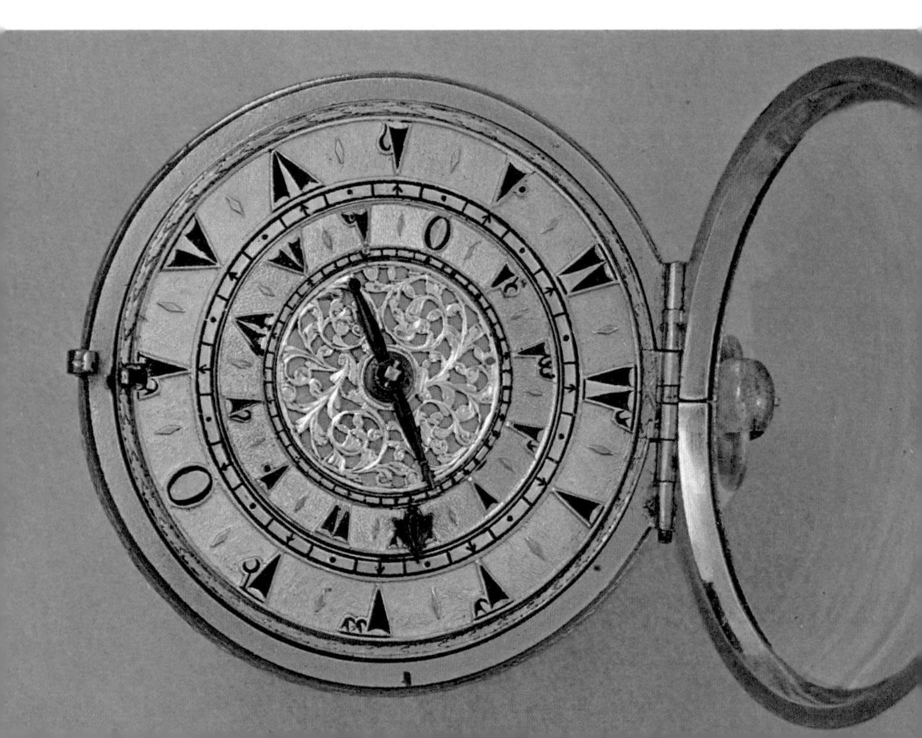

17 A French verge watch made by L. Vautyer in the 17th century. The case is enamelled with raised and pierced gold and enamel decoration.

18 A 17th century German pre-balance watch with fusee and chain. The case is made of rock crystal.

19 An English pair cased alarm watch made in the last third of the 17th century by Tompion.

20 A large French watch, *c.* 1680, by Fardoil of Paris. The silver case is pierced and engraved. The dial is made of gilt brass and the hand of blued steel.

21 The movement of the watch in Plate 20, showing the pendulum cock.

22 Although inscribed 'Barraud, London', this verge watch is Swiss. The sails of the windmill revolve. The pair cases are gilt metal set with paste and enamelled. The watch dates from 1756-94.

23 An English verge watch, dating from the late 17th century, with balance spring and wandering hours dial. It is inscribed 'Jos Windmills 0717'. The pair cases are silver, the outer being engraved with the arms of King William III.

24 An English verge watch, *c.* 1700, by Stammer, with a double six hour dial. This was an attempt to achieve an accurate indication of the minutes with only one hand.

25a An English double dialled watch, *c.* 1770, by David Pons.

25b *Opposite.* The 24 hour dial also carries the signs of the zodiac.

25c The movement has a cylinder escapement.

DEC 31 JAN 31 FEV 28 MAR 31 AVR 30 MAY 31 JUI 30 JUIL 31 AOU 31 SEP 30 OCT 31 NOV 30
60 5 10 15 20 25 30 35 40 45 50 55
CAI/176

26 Cased as a chronometer, this is nevertheless really a large and complicated astronomical watch. It is English, *c.* 1780 and was made by Margetts.

27 A Dutch watch, *c.* 1710. The beautiful gold repoussé case illustrates 'The Last Supper'.

28 This English gold watch case, dated 1774, is enamelled with a portrait by Henry Spicer.

29 An English spring detent chronometer. The gold case is hallmarked and dated 1780. The hands are also gold and the dial enamel. This watch has a half quarter dumb repeater.

30 A French cylinder 'souscription' watch made in a silver case, by the incomparable Breguet in 1798. An 'inexpensive' line, these watches were ordered and made in batches. The barrel is at the centre.

31 *Top*. An early enamel dial with cartouches. The case is piqué.

32 *Middle*. This watch is unusual as it bears a portrait inside the front cover.

33 *Bottom*. A superbly enamelled case.

34 This Swiss musical watch, dating from the early 19th century, strikes on five bells. The case is gold and has a beautifully pierced and engraved cuvette. The watch is decorated with split pearls and enamelled.

35 This is a Swiss quarter repeater watch with automata, dated 1800-25. Two cupids ring bells and at the bottom two figures operate a grinding wheel. The watch has a cylinder escapement and the case is gold.

36b *Left*. The back of the watch shown below. The gold case is decorated all over with grains, filigree and amethysts.

a *Right*. A Swiss verge watch, *c*. 1810. The movement is inscribed 'Fx. Pernetti à Geneve 13601'. The dial is gold with the numerals on oval plaques.

37 Breguet's testimony to Arnold, 1808. He fitted his 'tourbillon' with spring detent escapement to an Arnold watch.

38 Another Breguet watch, *c.* 1810, with one of his experimental escapements. It is a repeater with a vertical wheel natural lift escapement. Only the movement of this watch now exists.

39 A Swiss repeater watch with automata, made by Meuron and Co., 1800-25. The case is gold and the dial has coloured gold figures and foliage.

40 This club tooth lever watch, *c.* 1836, is inscribed 'Robert Roskell', but is in fact Swiss. The dial has a thermometer and a compass fitted into it.

41 This is a French cylinder watch with repeating dated 1812-25. The figures stike real bells.

42 Two cylinder watches. The top one is plain and delicate; the bottom one is decorated with enamelling and engraved and has a shaped case.

43 *Top Opposite.* This is a musical quarter striking clock watch, *c.* 1815, by Courvoisier.

44 *Bottom Opposite.* A French independent seconds watch, *c.* 1815.

45b *Left*. The back of the silver watch shown below. The large seconds beating balance nearly covers the top plate.

45a *Right*. This watch, *c*. 1815, is inscribed 'Ls George and Co., Berlin'. It is a quarter repeater with a Pouzait type lever escapement. The small upper dial shows the day and date.

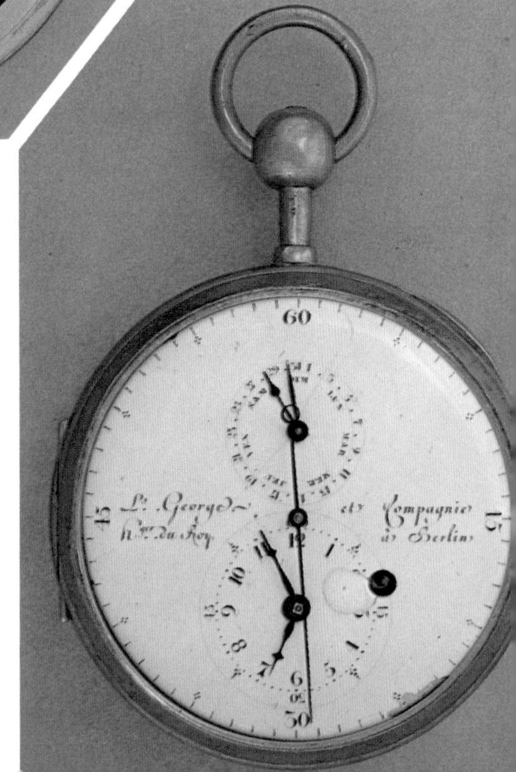

46 'A tact' gold watch by Breguet, *c.* 1820, with a cylinder escapement.

47 An English clock watch by French, Royal Exchange with a duplex escapement.

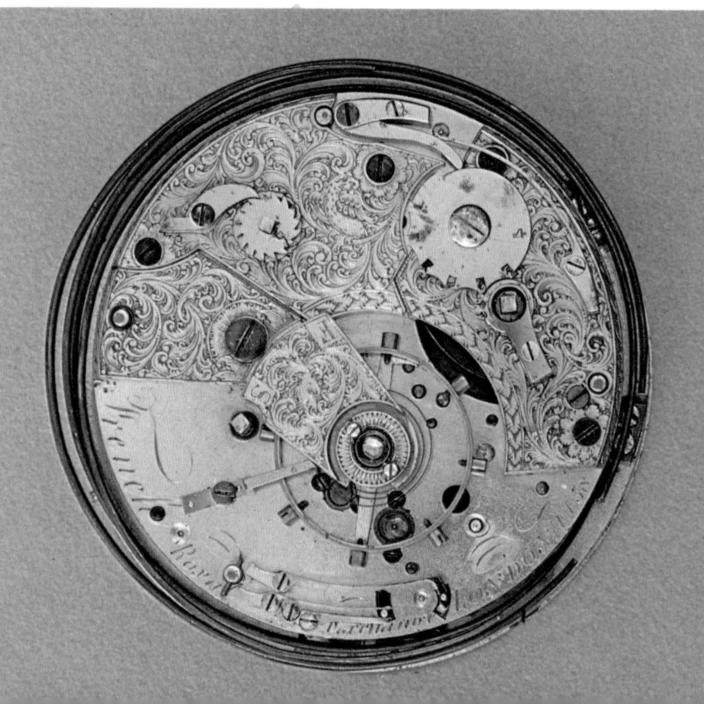

48 A very flat French cylinder watch. It has niello on gold decoration and digital display.

49 An English watch, *c.* 1820, with Massey type lever escapement and 'Liverpool jewelling'.

50 A Swiss watch, *c.* 1830, with Tavan pinwheel lever escapement and compensation curb.

51 An English watch, 1828-32, with a ratchet tooth lever escapement. It has a three arm bimetal balance.

52 A magnificent tourbillon with chronometer escapement. It is inscribed 'Hunt and Roskell, London', but is probably Swiss. The top illustration shows the movement.

54 A Viennese reproduction watch, *c.* 1840. It is a copy of a 1650 style watch. It has a crystal front and back with enamelled bands. The illustration below shows its single hand.

55 An English club tooth lever watch, *c.* 1846, by E. J. Dent. It is marked 'compound movement' and is an early example of a split seconds watch.

ball watch
:h an enam-
ed case and
ain.

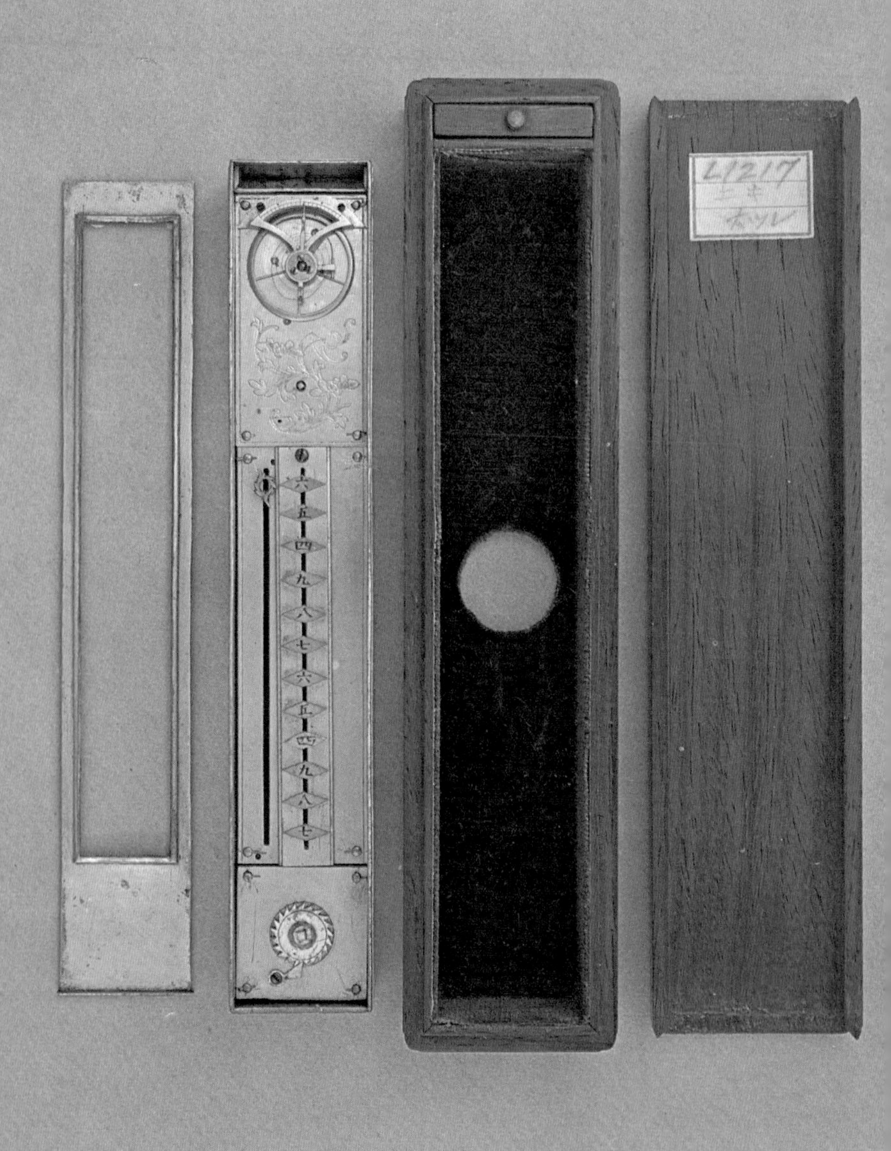

56 This is Japanese and is described as a paperweight or doctor's clock, but it is really a watch. The case is brass, while the outer case is made of shitan wood. It probably dates from the 19th century.

57 This is a Japanese inro watch of the mid-19th century. It has a verge escapement. The watch strikes and the bell can be seen through the aperture at the top. It was carried on the girdle.

58 This Japanese watch, with verge escapement, probably dates from the 19th century. It strikes on the Japanese system. The case is silver and the gilded metal dial has moveable steel characters.

59 The movement of this Swiss cylinder watch, *c.* 1860, is all made of steel. The style of the movement is Bovet. The dial is enamel and the case silver.

60 Often sold in pairs, these Swiss Bovet watches were decorated all over, both case and movement. This silver gilt case is enamelled and decorated with split pearls.

61 The movement of the watch shown opposite. It is enamelled and engraved. The escapement is Chinese duplex and the watch was made *c.* 1860.

62 The Swiss 'Roskopf' watch, *c.* 1870. This was the first watch to be made for the working man. It has a pin lever escapement and is keyless, with a rocking bar system.

63 Made by an American in Switzerland in 1876, the movement of this watch is inscribed 'Albert H. Potter and Co., Geneva'. The escapement is pivoted detent.

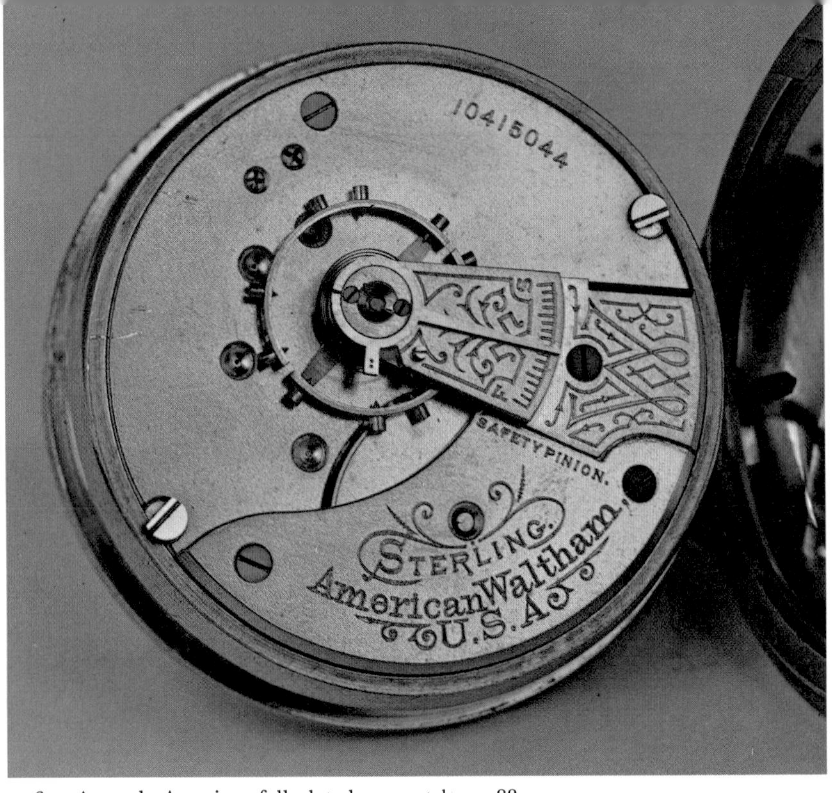

64 An early American full plate lever watch, *c.* 1880.

65 A Swiss one minute tourbillon with lever escapement, 1880.

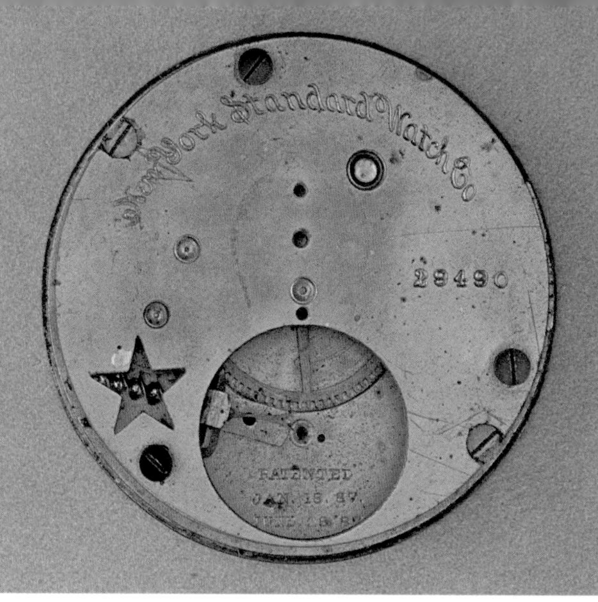

66 An American rarity, 1887. This watch has a worm gear in the train.

67 A German watch, *c.* 1880, with an English name.

68 This German movement was made by A. Lange and Son who also made the watch shown in Plate 67. It has a club tooth lever escapement and is a half quarter repeater and chronograph.

69 This American watch, *c.* 1888, by Waltham, has a crystal top plate and balance cock. It has a lever escapement. The case is later.

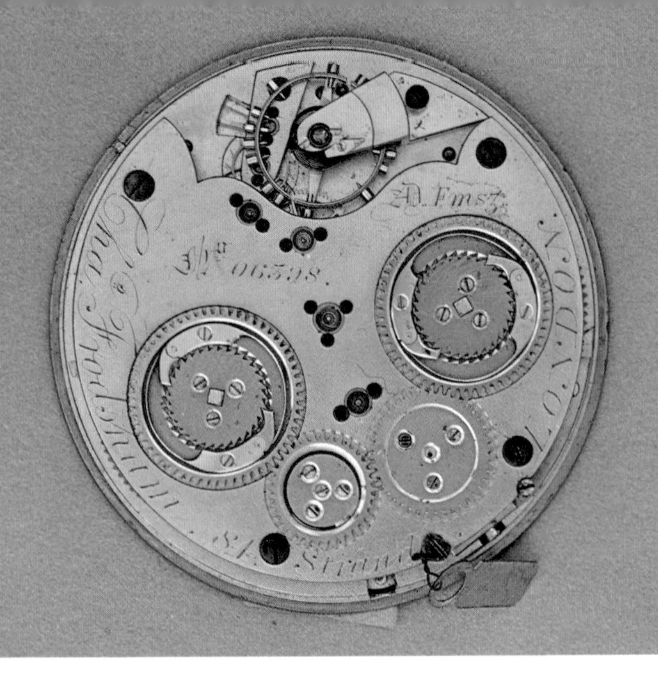

70 An English watch movement, *c.* 1890, by Frodsham. It is 8 day and has two barrels. The escapement is English lever.

71 A French movement, *c.* 1890. The bars make the word 'Paris'.

72 This Roskopf watch is later than that shown in Plate 62. It has acquired the form it was to take for decades.

73 A flat Swiss lever watch with engraved movement.

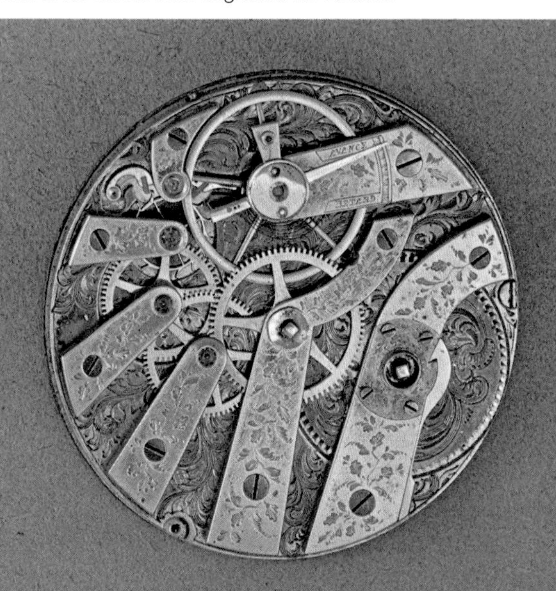

74 A Swiss tourbillon, *c.* 1920, by Fureur. It has a lever escapement and a going barrel. The case is silver.

75 This early 20th century English Karrusel watch is by Yeomans. It has centre seconds and a going barrel.

76 A superlative quality American watch, 1918, by Elgin. It is beautifully damascened and has up and down work.

77 An early 20th century tiny Swiss minute repeater. It measures 27mm in diameter.

78 A modern Swiss wrist watch, *c.* 1950, by Mathey-Tissot. It shows the day, date, month and phases of the moon and is a chronograph with a minute recorder.

79 The first successful self-winding wrist watch, *c.* 1930. It was designed by Harwood, an Englishman.

80 A self-winding watch prototype made by the author and P. W. Amis in 1957. It has a shock-protected weight.

81 The first successful electric watch, 1957, made in the U.S.A. by the Hamilton Watch Co.

82 *Above.* The first truly electronic watch. The balance and spring has given way to the tuning fork. American made, it was designed by a Swiss in 1963.

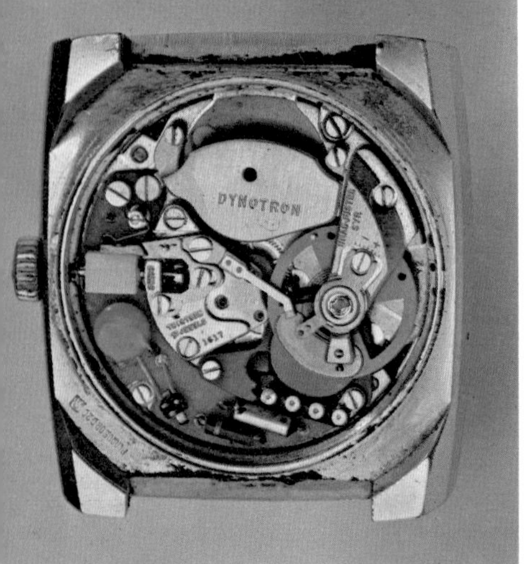

83 *Above.* This Swiss watch, made in 1967, was one of the last battery watches with a balance and spring. There were no contacts.

84a *Below.* A remarkable design feat – a self-winding chronograph with date work. Made by the Hamilton Watch Co. in 1970.

84b *Below.* The chronograph work of the watch above can be lifted off as a complete unit, thus facilitating repair.

5 and 86 (*over*). These are modern examples of colour used in watches. These matching watches, earrings and rings were made by the prestigious firm of Patek Philippe of Geneva in 1977. Those above are decorated with diamonds, onyx and malachite. Those shown over the page have diamonds and lapis lazuli.

PATEK PHILIPPE
GENEVE

5 The Decoration of the Case

These are by: (1) engraving, (2) chiselling, (3) casting, (4) repoussé work, (5) engine turning, (6) presswork, (7) piercing.

Of course, more than one of these techniques could be employed at the same time and frequently were. Surfaces could also be enamelled although this will be dealt with separately. Cases could also be covered in fish skin, tortoise-shell, inlaid as in damascening work and, of course, set with precious and semi-precious stones.

The movement itself can also be treated in many of the above ways although decoration is found less frequently than on the case. The movement could also be damascened, a process which has nothing to do with the inlaying type of damascening but consists of a type of engine turning. This was frequently used on the better later American watches (see Plate 76).

To return to watch cases, however, some were first cast and then worked upon by chiselling to crisp up the relief and also by engraving for fine detail (see Fig. 5). Chiselling is done with a hammer and punch whereas engraving is done with a graver which is pushed by hand. Many shapes of graver are used to produce different effects. Pierced work can also be cast in. Chiselling and engraving are often combined as in champlevé engraving, where the background is chiselled back to a ground level and then engraved with fine grooves, or some other pattern, to give it a dark appearance. The foreground can also be finished by engraving. The pockets so formed can be enamelled or filled with niello. Repoussé, an ancient art, is done by punching up the pattern from behind. (See Plate 27. The dial of the watch in this repoussé case is shown in Fig. 11.)

Fig. 11 Typical Dutch dial. This is the front of the watch whose outer case is shown in Plate 27.

Enamelled Cases

Enamelled cases are a subject on their own, and of course appeal to many people who have no thought at all as to the mechanism they house. Enamel is a glass composed of silica, red lead and potash. When used without colouring it is a transparent flux called 'fondant', but it is usual for it to be coloured. It is coloured variously by the addition of the appropriate metal oxide and may be hard or soft according to its composition. Hard enamel can only be applied to metals with a high melting point but retains its surface and colour indefinitely. It can, however, be cracked and chipped easily. Soft enamel which does not crack or chip is unfortunately easily scratched.

Enamel decoration takes six forms:

1. Plain enamel (fondant).
2. Painting on enamel.
3. Painting in enamel.
4. Basse-taille.
5. Cloisonné.
6. Champlevé.

1. Fondant is hardly ever found on its own.

2. Painting on enamel was evolved in about 1630 by Jean Tontin, a French goldsmith (1578–1644). A plain enamel ground is laid down and the painting done with coloured enamels which are then fired. The whole is then covered with another layer of fondant and again fired. This is also the process used in making enamel dials.

The early painted enamel cases were splendid, but after about 1660 the quality deteriorated and colours were less brilliant. The favourite subject then became women in various stages of undress being ogled by old men or satyrs. The Haut family were the most renowned artists of this style (see Fig. 12). A magnificent example of

Fig. 12 Enamelling in the style of the Haut Family.

English enamelling is shown in Plate 28; the signatures inside this case are shown in Fig. 13.

Fig. 13 Signatures inside the case shown in Plate 28.

The Geneva school continued throughout the 18th century, while the Blois school declined. Latterly Swiss artists have specialised in scenes known as 'gallant'. Not to put too fine a point on it they are pornographic and were often associated with the automata already mentioned.

3. Painting in enamel was developed during the 16th century, after which it is seldom found. It found its simplest expression on watch cases and some of these are the most beautiful enamelled cases to be found. Small blobs of enamel are deposited on the case, usually in a floral pattern. After firing the blobs remain in relief (see Plate 8).

4. Basse-taille is a layer of transparent coloured enamel over a chiselled or engine turned gold ground. Some cases in which the enamelling is over chiselling survive from before 1650, but in the main basse-taille did not become fashionable until after 1780. Most surviving examples are Swiss. (See Fig. 14 which is the back of the watch shown in Plate 34.)

5. Cloisonné is enamelling in which the outline of the pattern is formed of thin wires, usually gold, soldered to the metal ground

and then filled with coloured enamels. The whole is then ground and polished to a uniform surface.

6. Champlevé, the earliest form of enamelling, is most frequently met with and was used extensively on dials as well as cases. Here the pattern is hollowed out and then filled with enamel.

Fig. 14a The back of the watch shown in Plate 34.

Fig. 14b A view of the back of the watch shown in Plate 34. Note the beautiful cuvette.

Fig. 14c Dial of watch shown in Plate 34.

Niello is a specialised form of champlevé most frequently done on silver but also on gold. Here the filling is black, this being an alloy of silver, lead, copper and sulphides (see Plate 48).

Under certain circumstances champlevé may be difficult to distinguish from cloisonné. It is interesting that brass and copper need to be enamelled on both sides or the metal would distort in the firing process. Gold is the only metal that need only be enamelled on the one side.

Rock Crystal Cases

Early examples of rock crystal cases are to be found such as the case reproduced in Plate 54. Usually the case is facetted, but not invariably. The case consists of two parts, the main part, which is hollowed out of a single piece of crystal, and the lid. Usually each part was mounted in a bezel and these hinged together. In the 17th century, the cases of small watches were hollowed out of precious or semi-precious stones such as amethyst. Again, cases were set with slabs of cornelian or agate.

The work of lapidaries was seldom used in the 18th century but during the early 19th century Swiss watches were often decorated with brilliants or split pearls, often associated with enamelling (see Fig. 15).

Having dwelt at some length on the decorative aspects of watches, some technical points responsible for these aspects have been noted. However, the main purpose of a watch is to tell the time and we must now take up the story of the changes that transformed the crude and inaccurate instrument that was the watch of the 17th century into the accurate machine that it had become by the end of the 18th century.

Fig. 15 The front of the watch shown in Plates 60 and 61.

6 Mechanical Considerations

The Watch after the invention of the Balance Spring

If a spring is used to control the beats of a balance, then in theory at least, such an assembly is isochronous, that is, it will perform a swing of a large arc in the same time as a swing of a small arc. Therefore, it will not matter if the source of power is not uniform. It was 1675 before this was realised, apparently independently, by Christian Huyghens, the Dutch Mathematician, and the London Physicist, Dr. Robert Hooke.

However, although theory suggested that the balance spring could make timekeeping perfect, this was unfortunately only under mathematical and not true-life conditions. In fact, although the balance spring enabled a radical improvement in timekeeping to take place, it revealed the shortcomings of the verge escapement in no uncertain fashion. For the best results, the loss in balance swing due to friction of the pivots and the fanning of the air, should be made good by an impulse of zero duration given exactly at the mid-point of the balance swing. This is when all the energy of the moving system is stored in the balance momentum and not as stress in the spring. For the rest of the swing of the balance it should be undisturbed. The results of any disturbance on the timekeeping abilities of the balance and spring become worse, as they occur nearer to the point of maximum swing. The verge escapement interferes with the balance for all of its swing and the conditions at maximum swing are especially bad, since at this time the balance is forcing the escape wheel to recoil against the full force of the motive power.

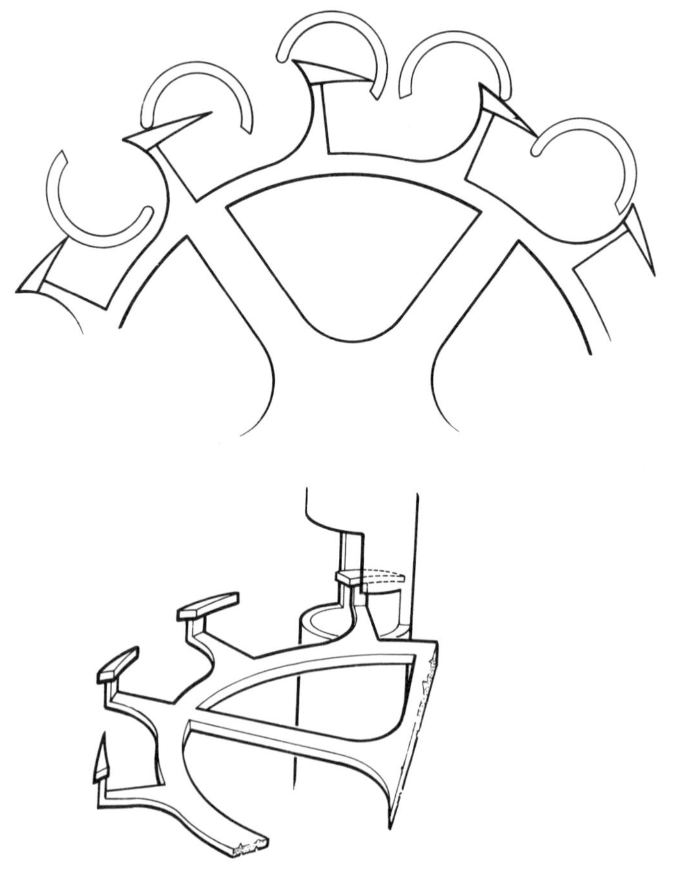

Fig. 16 Cylinder escapement.

The Cylinder Escapement

Around 1726 an effective improvement on the verge, the cylinder escapement was produced by George Graham. It is shown in Fig. 16. The cylinder escapement is a 'dead beat' escapement, that is there is no recoil of the escape wheel during the escapement action. It is fitting that Graham should have designed this escapement, for he also designed the dead beat escapement for clocks which reigned supreme for two centuries as the escapement for precision clocks.

The cylinder escapement is what is known as a frictional rest escapement, that is, the escape wheel teeth are in contact with the balance during supplementary arc. (Supplementary arc is that portion of the arc over and above that minimum required for the action of impulse to take place.) The best escapements had wheels of steel and cylinders of ruby.

The reason why the cylinder escapement was an improvement on the verge escapement is that no longer did the balance have to recoil the train at the end of its swing. It is true that the balance was still in contact with the escapement for the whole of its swing, but under much improved conditions. The impulse was given over a much smaller angle of travel of the balance, therefore getting a little nearer to the ideal of a sharp blow about the zero position. At any other time, the only disturbance was caused by the friction between the escape wheel tooth tip and the polished surfaces of the cylinder.

Not only was the cylinder escapement an improvement on the verge but it also enabled slimmer movements to be made. The escape wheel of the cylinder escapement was placed in the same plane as the plates and not at right angles to them as in the verge. As a result, the cylinder escapement was called the horizontal escapement when Graham introduced it. After 1726, Graham used the cylinder escapement almost exclusively and following him Thomas Mudge used it frequently. Thomas Mudge was Graham's apprentice and was renowned for the beauty of his workmanship and his inventiveness.

About 150 years before the invention of the balance spring it was realised that if a ship carried an accurate timekeeper, then a comparison of local time, (as found by observation), with a standard time at a known place could establish how far a ship was east or west of this place. Since it was already possible to determine the latitude of a ship by observation, this additional determination of longitude would pin-point the position of a ship on the face of the globe. When the pendulum was first invented, there was hope that this might prove to be the answer, but the problem of insulating the pendulum from the influence of the motion of a ship at sea was too great tobe overcome at that time.

Finding the position of a ship at sea was of vital importance, since any long voyage incurred risks of shipwreck that were so high, that the safe arrival of any ship was a mere gamble. Fortunes were lost every year and it was a matter of great concern to all the major maritime nations. So much so that in 1714, on the advice of Isaac Newton, the British Government offered a prize of £20,000 to any person who could produce a time-keeping device that would be accurate and remain accurate at sea. This was a vast sum of money worth probably a million pounds sterling today, and it inspired some of the finest clock and watchmakers of the day to compete. It inspired a lot of other aspirants too, so many that a Board of Longtitude was set up to consider their many and varied claims.

The maximum prize was only to be awarded if the timekeeper could ascertain longitude to half a degree. Since this was over a period of six weeks, it could mean an accrued error of no more than two minutes in six weeks, or an average of about three seconds a day! When one realises that the average watch of the time had errors of minutes a day it is evident that something radical had to occur if a portable timekeeper was to be devised that could keep time to three seconds a day.

John Harrison, after three attempts with clocks eventually succeeded in winning the prize with what was in fact a large watch. He modified the verge escapement (see Fig. 17), fitted a constant force device in the gear train to eliminate errors from the mainspring, fusee and gears, and solved the problem of temperature compensation. He also used jewels in his movement. This was possible because of the invention in 1704 of a method of piercing and cutting stones, so that they could be used as pivot bearings in clocks and watches. Such jewel bearings eventually appeared in English watches as early as 1750, but the great A. L. Breguet appears to have been the first to use jewel

Fig. 17 John Harrison's modified verge escapement.

bearings on the Continent. The purpose of jewels was, of course, to reduce friction and wear, and without them no fine adjustments would be of a lasting nature.

Regulating the Balance and Spring and Temperature Compensation

If the length of the balance spring is changed, then the timekeeping of the balance and spring assembly also changes. Making the spring longer makes the watch go slow, making it shorter, fast. The easiest way of effectively changing the length of the balance spring was to pass it between pins that were near to its outer end, and make these pins moveable. By using this method the old way of regulating the watch by altering the set-up of the mainspring was discontinued.

Barrow is reputed to be the first to invent the balance spring regulator. He took the worm that was once used to alter the mainspring set-up and instead made it move the pins that embraced the balance spring. For this purpose, the part of the balance spring traversed by the pins must be straight, otherwise the body of the spring would be moved as the pins traversed it.

Tompion invented a type of regulator whose pins moved concentrically with the spring centre, so that they more nearly followed the natural spiral of the spring. The pins were carried by a geared segment driven by a pinion, that carried a square for the winding key. The pinion was covered by a divided dial to assist in making accurate changes. This form of regulator omitted the balance spring that lay beneath the balance.

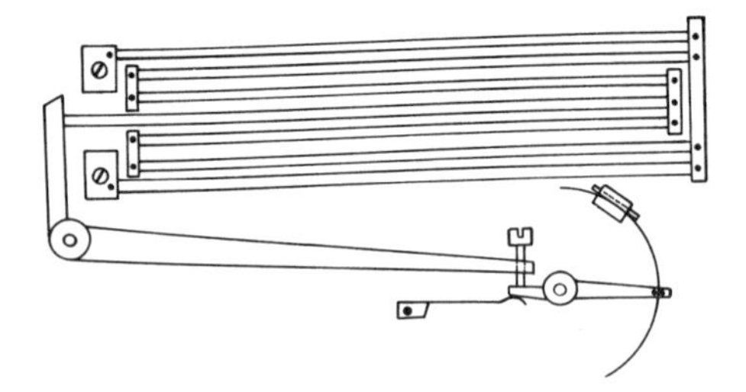

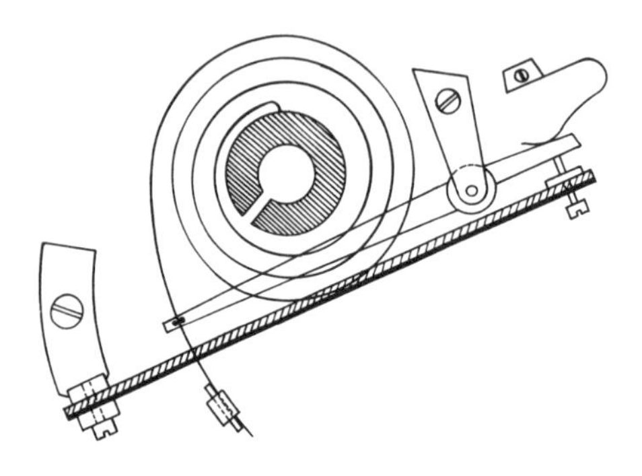

Fig. 18 Top: *Berthoud's type of temperature compensation using a 'grid iron'.* Bottom: *temperature compensation with single bimetal strip.*

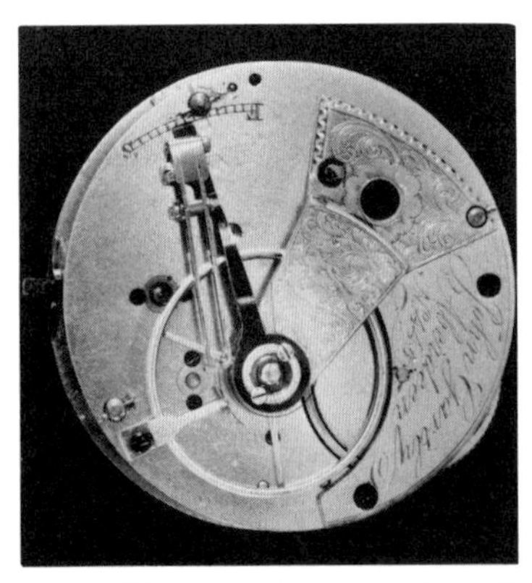

Fig. 19 'Sugar tongs' compensation curb.

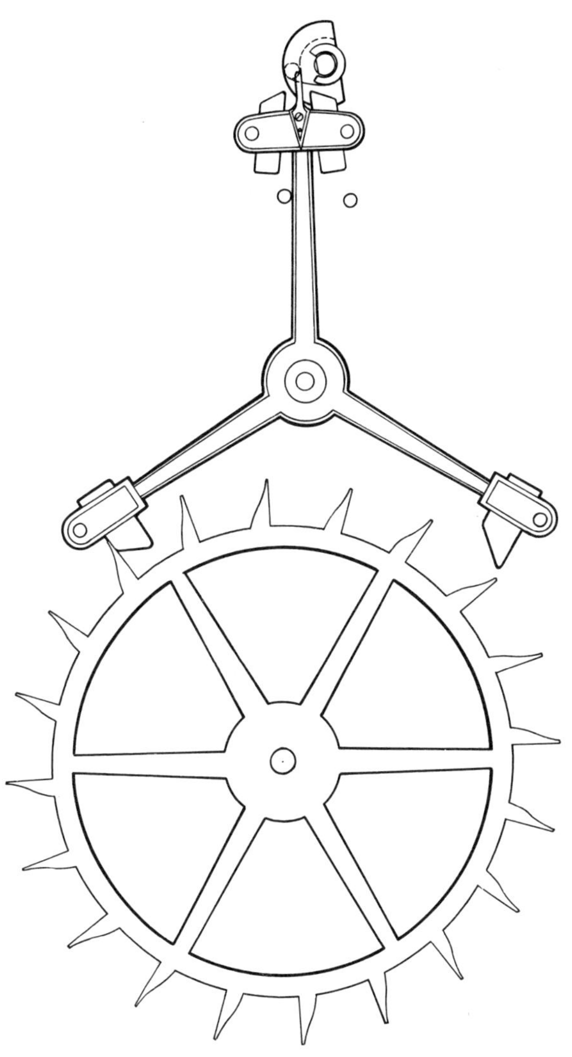

Fig. 20 Mudge's lever escapement laid out in a straight line.

Maintaining Power

During winding early watches (and clocks) in the main stopped since the act of winding robbed the train of its power. The watch was not then accurate enough for this to worry anyone. Once the way was clear to improve the timekeeping of a watch sufficiently it became important for the watch not to stop each time it was wound. John Harrison invented a device called maintaining power that kept the watch going whilst it was wound and was so good that it is still in use today.

Temperature Compensation

If a brass balance has a steel balance spring and no form of temperature compensation it will display timekeeping changes of *six minutes a day* for a change in temperature from 30°F (— 1·1°C) to 90°F (32·2°C). Harrison solved the problem of these changes by using a compensation curb. If these pins can also be moved by a compensation device whose position changes with changes of temperature, then when this device is correctly adjusted the rate of the watch can be made to stay reasonably constant, although temperature changes occur. An early and a later type of compensation device is shown in Fig. 18.

Harrison's curb pins were moved by bimetallic strips made from brass and steel riveted together. The differing coefficients of expansion of the two metals cause a strip constructed of them to take on more or less of a curve with changes of temperature. The portion of the spring along which the curb pins move has to be shaped to fit in with the way in which these pins move. Another version known as the 'sugar tongs' has limbs that affect the separation (see Fig. 19).

Mudge's Lever Escapement

Thomas Mudge, already mentioned, one of the great men of horology, produced in the middle of the 18th century, an escapement that was to be the single most important escapement ever, the lever escapement. Figure 20 shows the escapement in detail, although it is rearranged to show how closely it resembles the English lever escapement which held favour for so many years and which appeared in what are probably some of the finest watches ever to be made.

7 The Emergence of the Precision Watch

Despite the discoveries of both Harrison and Mudge no real improvements had occurred in the common watch by the third quarter of the 18th century. They were still almost identical to the watches produced by Graham fifty years earlier. However, there were exceptions to this – Breguet in France and Earnshaw and Arnold in England.

Breguet was probably the most brilliant horologist of his and indeed any time. He had extraordinary ability and some of his improvements are still in use today. Breguet established his business in 1775. At that time John Arnold was probably the best-known watchmaker in Europe and Breguet and Arnold enjoyed a firm friendship, cemented during Breguet's visits to London. So great was their mutual regard that each sent the other his son for instruction. Plate 37 shows a watch that commemorates their friendship.

John Arnold was established by 1764 and in this year presented to George III a watch with ruby cylinder and repeating mechanism. Its size was little more than ½ in (13 mm) in diameter. Such a watch would be an item of note even today; at the time it marked Arnold as a craftsman of exceptional ability. Added to this he was a man of great intelligence, who appreciated the fundamentals of horology.

John Arnold was responsible for the helical balance spring that has been used in almost every marine chronometer from that day to this. He patented this spring in 1776 and Plate 29 shows a watch made in 1779 containing such a spring. At the same time he patented a balance, which itself incorporated the temperature compensation device, thus setting the pattern for two centuries. In the previous year

John Harrison had suggested the temperature compensation balance as an alternative to the compensation curb; also Le Roy in France had made one.

Whether or not Arnold takes the full credit for the chronometer or detent escapement is arguable, since Ferdinand Berthoud devised a spring detent escapement and Thomas Earnshaw was a counter claimant. Whether this is so or not seems relatively unimportant now. Certainly, the escapement is the most beautiful in its simplicity and elegance and approaches the ideal. It is shown in Figs. 21a and 21b.

In one grand step the escapement moved from being a relatively crude to a very sophisticated device. For the first time, if one ignores Mudge's lever escapement, the balance was free for the major part of its swing, and impulse and unlocking occurred over a relatively small angle.

The results were immediate, Arnold made a watch which throughout a year of testing kept within an error of three seconds a day. But, Arnold's escapement, however good, was still inferior to Earnshaw's, an escapement similar in principle but having the added attribute

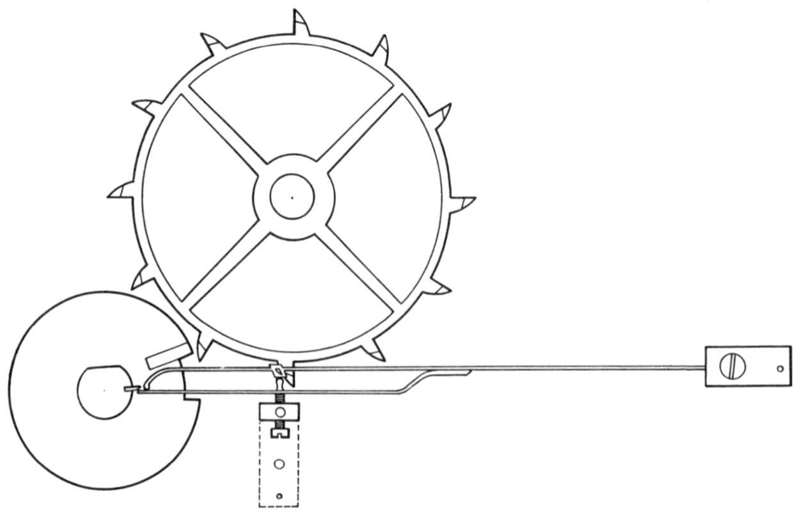

Fig. 21a Arnold's spring detent escapement. (Escape wheel revolves clockwise in the drawing.)

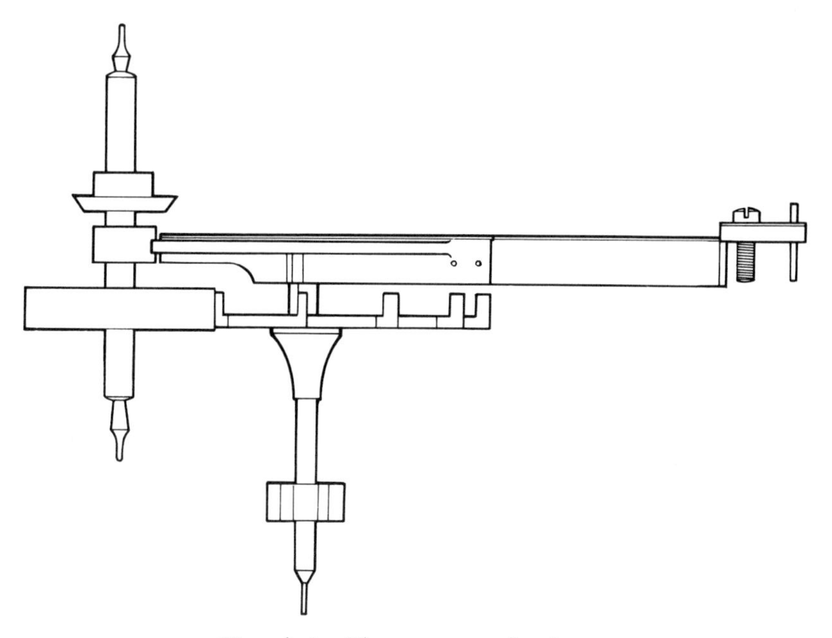

Fig. 21b Arnold's escapement – edge view.

that it did not require oiling at the escape wheel teeth (see Figs. 22a and 22b).

Arnold also improved his helical balance springs by curving the ends into a smaller radius than the main body of the spring. By correctly curving these spring ends, Arnold was able to both reduce the effects of the side thrusts on the balance pivots consequent upon the uneven dilation of the spring, and also to utilise a remaining thrust to minimise the errors due to the escapement.

Thus in Arnold's lifetime an enormous leap forward was taken in producing accurate and reliable watches, simple and elegant in design and construction. This also meant that they could be made in quantity and at a price that ensured their proliferation. This enabled the Royal Navy, upon which England depended for her survival, to navigate accurately anywhere on the face of the globe. The contribution this made to both England's supremacy in horology and her accumulating wealth is difficult to overestimate.

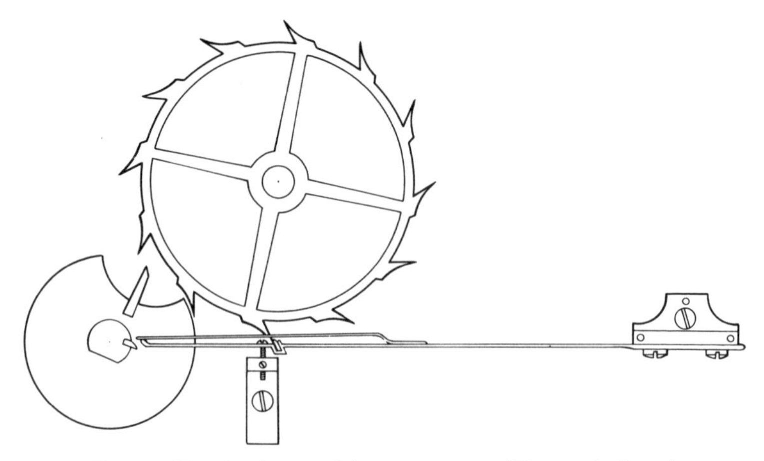

Fig. 22a Earnshaw's type of detent escapement. (Escape wheel revolves anti-clockwise in the drawing.)

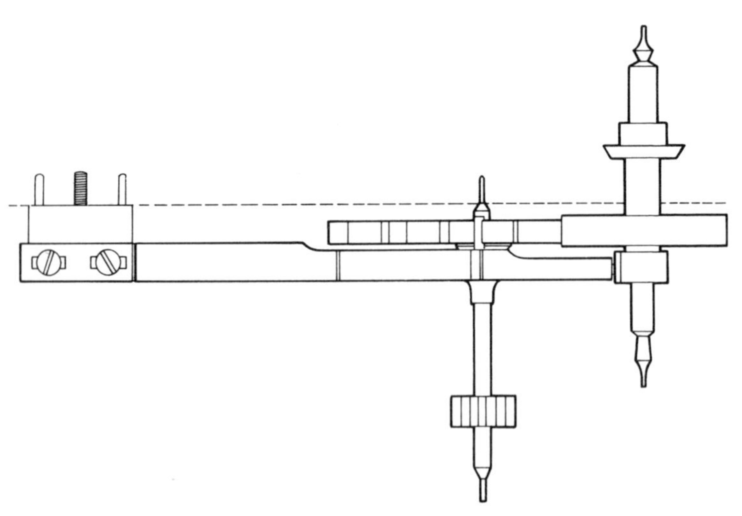

Fig. 22b Earnshaw's escapement from the edge.

The Temperature Compensation Balance

Although the devices that affected temperature compensation of the balance and spring assembly by moving the curb pins were to continue to be used for many years, the compensation balance was the proper answer to the problem. Many and varied were the types that were used.

Many of these need not be discussed here, since they were only used in marine chronometers. The main type of balance used has two, three or four arms made of bimetallic strips. These carried weights or screws, which could be moved nearer to or further away from the free end of the bimetallic strips, so as to afford a degree of adjustment. Early watches often had segment weights, but screws eventually became almost universal. Sometimes, even when all of the holes at the free end of the strip were full of screws, the degree of compensation was not sufficient. As a result, one sometimes sees the brass screws at the end replaced by gold or even platinum screws. When extremely accurate watches are being made, that are to be compensated for middle temperature error, the Guillaume balance is used with a steel spring.

This will be discussed in the section on Guillaume as will the so-called 'self compensating balance spring'.

8 *Breguet As Mechanic*

Abraham Louis Breguet, mentioned already, was probably the most remarkable horologist ever to live. His vast range of accomplishments could have taken up this entire book. All of his work had a distinctive character and often contained technical innovations.

His inventions include the following:

1. The tourbillon, where the entire balance and escapement revolves to eliminate the positional errors that occur when the watch is on edge.
2. A Montre 'à tact', which is a watch with a touch piece at the edge of the case which indicates the time. A hand at the back of the case is turned, until it is felt to contact the internal mechanism, when its position can be felt relative to the hour pieces. This meant that the time could be ascertained in the dark or by a blind person.
3. Shock-proof bearings for the balance pivots called 'parachute'.
4. A Ratchet winding key to prevent attempts to wind in the wrong direction.

There were also many improvements relating to clocks, to say nothing of a marriage between watches and clocks, where the clock sets the watch to time and regulates it in accordance with the errors so corrected. In later developments of this, the clock winds the watch each day and sets it to time, although no regulation is done.

When Breguet came on the scene on the Continent little had happened there as regards the watch for fifty years. Jean Antoine Lepine, it is true, had invented an entirely new design of watch. He was born in 1720 at Challex, France, a district near to Geneva. This district supplied ebauche to Geneva during the 18th century. At the age of twenty-four Lepine left for Paris, where he soon made his mark. In his design he discarded the fusee and used the going barrel. The top plate and its supporting pillars were also discarded and replaced by separate bridges and cocks. This meant that thinner calibres could be made and after several improvements the manufacture of the Lepine calibre expanded greatly from 1795 onwards. Lepine used both the cylinder and the virgule escapement. Some early development work had been done by Ferdinand Berthoud and Le Roy to produce a precision timekeeper but this had had no impact upon the common watch.

In England everyone was well catered for. Arnold and Earnshaw were producing precision watches but most French watches were still using the verge and the virgule. Neither can keep good time, although the verge has the merit of being reliable and needing little attention for long periods.

No individual makers had given serious thought to improving the timekeeping qualities of the ordinary watch and the precision watch was still virtually unknown on the Continent. In England, it is true the industry had been given an enormous fillip by the gigantic prize offered by the government to produce an accurate portable timekeeper. Only minor prizes were offered in France by the Académie des Sciences out of an endowment left them by De Meslay.

Someone was needed to inspire the French industry to make the contribution that their position in the horological world merited and this task was left to Breguet. A man of genius, imagination and industry, nothing was ever to be quite the same again after him.

The movements of his early watches give little indication, however, of what was to follow, although he did give up the verge in favour of the cylinder escapement. However, he was to spend thirty years in his search for the perfection in his escapements. This is one reason for the variety of his work. The other, is his evident love of complexity. His Marie Antoinette watch was a triumph of the time (see pages 150 and 151).

However, Breguet's achievements were not only in the field of technical innovation, for he brought his own unmistakeable brand of elegance to the current Continental vogue for flat watches. This desire for less bulky and more elegant watches was virtually ignored by the English. Technical excellence was the criterion that governed the designs of the best English makers, who saw no need to sacrifice their principles to the fashions on the Continent. However, this was really the beginning of that inflexibility that was ultimately to destroy the English watchmaking industry, for in the end, the advances made on all fronts in horology made the flatter mass-produced watch a truly excellent product.

9 The Lever and Duplex Escapements

Although the lever was eventually to become the dominant watch escapement, its merits were generally unrecognised during the time that the chronometer was being developed. This is understandable since it is, in fact, a more difficult escapement to manufacture and its best geometry had not been established.

However, Mudge's patron, Count Von Bruhle, was very keen to have some watches made incorporating Mudge's escapement and managed to persuade Josiah Emery to undertake the task. Emery, a Swiss-born watchmaker, resident in London, was a very able man but was reluctant to start the work, saying it was too difficult.

Nevertheless, he produced his first lever watch in 1782, although his lever escapement was not just a copy of Mudge's. He incorporated improvements relating to the proportions and also employed a compensation balance. This watch proved superior in wear, both to Mudge's original watch and to the chronometer in general. Although not a superior timekeeper to the chronometer, it was not so inclined to stop during wear.

Emery continued to make these watches making nearly forty until his death in 1797. He continued to make improvements in the escapement, but these related more to ease of manufacture than to fundamental improvements.

Others used the lever escapement, such as Grant, Pendleton, Perigal and Dutton, but none incorporated what is called 'draw'. Without draw, the lever tended to rub on the balance roller when the watch was moved roughly or knocked. Alterations in the geometry of the pallets could overcome this problem, by biasing the lever towards the stops that limited its movement. Even though this was appreciated by many makers, they were reluctant to incorporate the change, since the force that kept the lever against the banking pins had to be overcome each time unlocking occurred, thus wasting some of the available power.

It was a man named John Leroux, who first introduced a recoil to the escape wheel as unlocking occurred, the recoil that thus gave draw.

The Duplex Escapement

However, about this time (1782) the duplex escapement was introduced into England and this was taken up as the alternative escapement to the chronometer for the precision watch (see Fig. 23).

This escapement was probably French, but its origins are obscure, its invention having been claimed by Robert Hooke, Pierre Le Roy, J. B. Dutertre, Thomas Tryer and others. Hooke may have possibly had the basic idea, but it is probable that J. B. Dutertre devised it in its early form in about 1725, although Le Roy perfected it in about 1750. The French makers dropped it because it really needed a fusee to give good results. Thomas Tryer took out a patent for it in England in 1782 and for some time it was known as Tryer's escapement. As in the chronometer, the escape wheel delivers impulse directly to the balance. One set of teeth perform this function whilst another set of teeth perform the locking function. These locking teeth are long pointed teeth, which intersect with the balance staff or a small roller mounted upon it and pass through a notch in the same when impulse is to be delivered. Thus the escapement is a frictional rest type, that is, the wheel is always rubbing on the balance. It is theoretically superior to the lever escapement without draw, in that the rubbing is consistent instead of intermittent and variable.

Furthermore, like the chronometer, the duplex escapement does not have to be oiled at the impulse teeth, and because of this it was considered for some time to be a serious rival to the chronometer. However, it had to be made with great care to function reliably for long periods and still needs oil at the locking teeth.

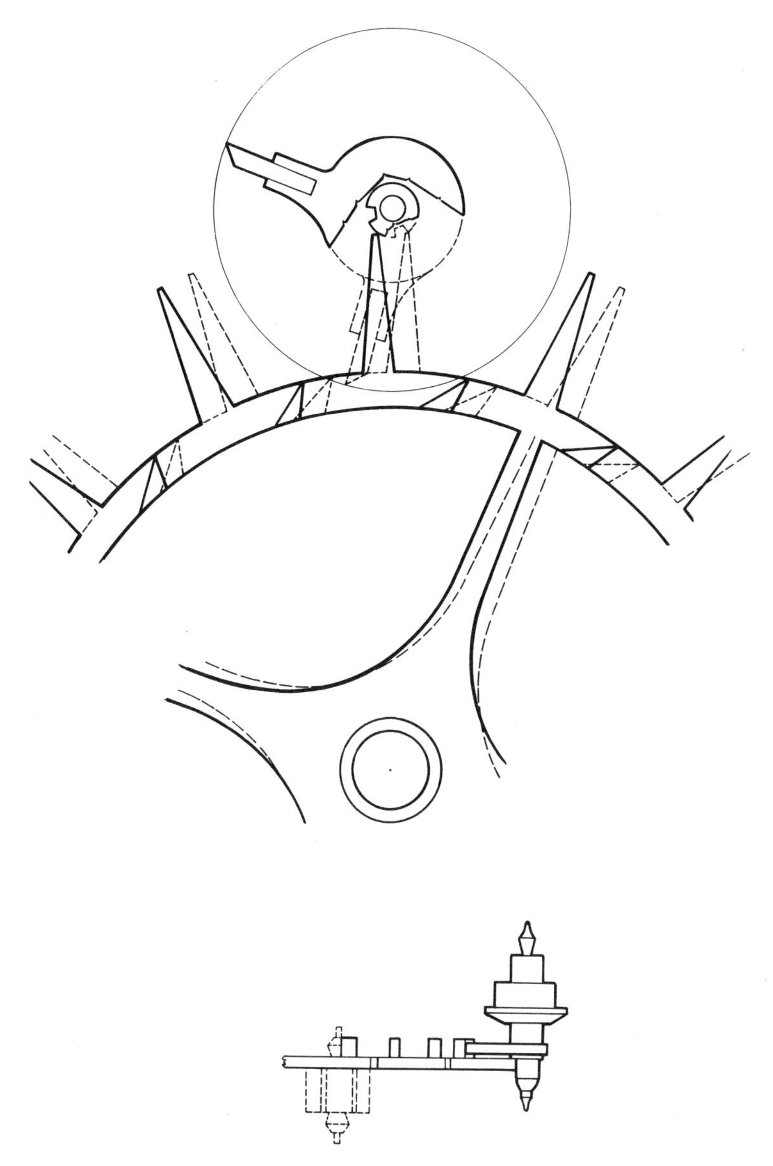

Fig. 23 Duplex escapement. (Wheel revolves clockwise in the drawing.)

The Lever Escapement Part II

As time passed the influence of Breguet's work on the Continent began to affect styles, even in English watches, and watches became thinner. For technical reasons, the duplex was not as suitable for thin watches as the lever and the development of the rack lever, in an improved form by Peter Litherland in 1792, helped at last to get the lever escapement under way (see Fig. 24).

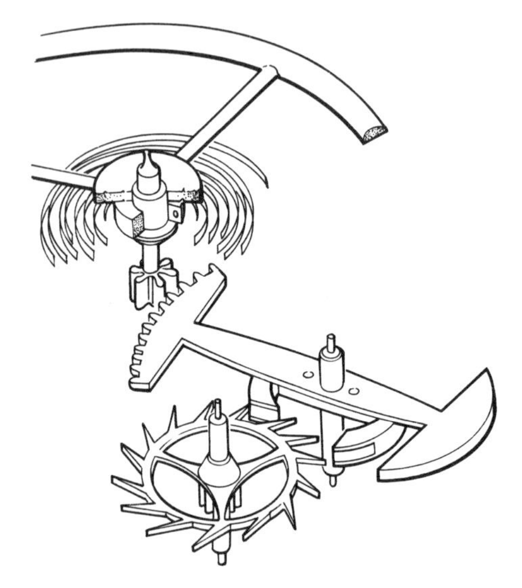

Fig. 24 Rack lever escapement.

No safety action was needed for this escapement, since the connection between the lever and the balance was affected by gearing. The lever end carried a segment of a gear, and a pinion was incorporated into the balance staff. Since these gears were always in mesh, no derangement was possible.

Tools were made by the industry to overcome the difficulties of manufacture, and many Liverpool makers began to turn out rack lever watches on a semi-mass production basis. In 1815 Edward Massey, a Liverpool maker, introduced a detached version of the escapement which, after fairly minor variations, became the English lever escapement for the remainder of the 19th century.

Breguet had also been working on the lever escapement on the Continent, often including incredible refinements. However, he had a curious 'blind spot' with regard to the necessity for draw in the escapement and still made no provision for this necessary modification, fully accepted by English makers by 1820.

In its final form the lever escapement was to approach the chronometer in accuracy and to exceed it in reliability where watches were concerned.

The Club-tooth Lever

The English lever escapement was characterised by the feature that all of the lift was on the pallets. This meant that only the tip of the wheel tooth slid along the pallet impulse face to give impulse to the lever (see Fig. 25).

However, at an early stage experiments had been made with a lever escapement that had divided lift, that is with half the impulse face on the pallets and half on the escape-wheel teeth. This meant that the tooth did not terminate with a tip but with a suitably angled impulse plane. First, the corner of the wheel tooth slid down the pallet impulse face, then, when it reached the end of the pallet face, the corner of the pallet slid along the escape-wheel tooth impulse face. The benefits of this arrangement were twofold. Firstly, the thickness of the tooth tip is utilised to give impulse action and does not represent wasted motion, as in an escapement with all the lift on the pallets. Secondly, the oil tended to remain where it should be for a longer period.

Although first designed in England the club-toothed lever (see Fig. 26) did not find much favour with English makers, although the Dent watch in Plate 55 had this type of escapement. It came into more general use in the 20th century, but this coincided with the death of the English watchmaking industry and it was left to the Swiss and the Americans to make it almost the universal escapement. The best of the late Swiss watches had escapements that were paired to the bone to achieve lightness. The escape wheel was recessed, so that only the boss and the tips of the teeth were at full thickness. The pallets were very delicate too. There was a phase when poising pieces were added

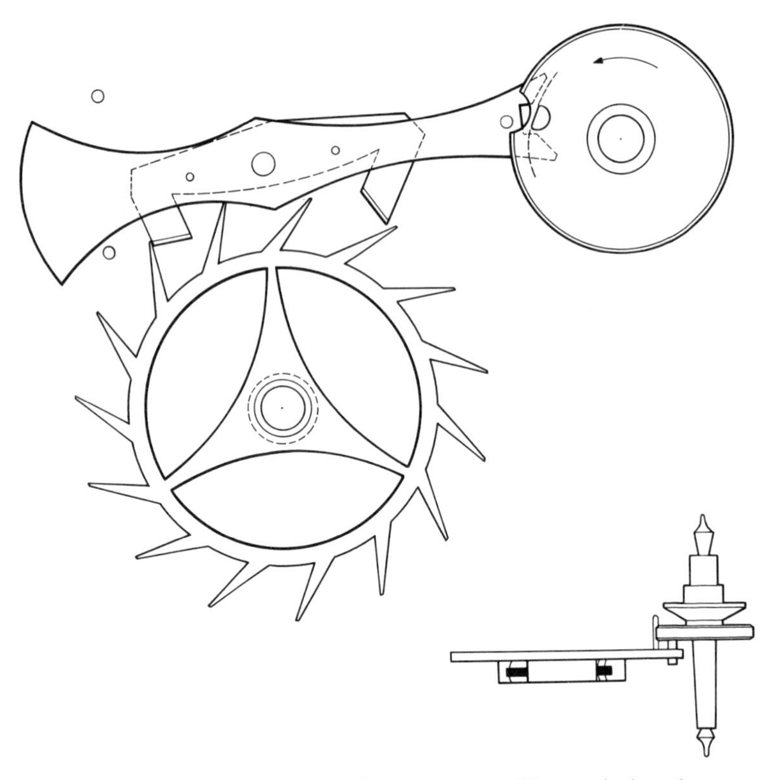

Fig. 25 English or ratchet tooth lever escapement. (Escape wheel revolves clockwise in the drawing.) In the edge view the dark portions are jewels.

to the pallets, in the mistaken belief that it was more important to try to poise the pallets than to make them as light as possible. For the same reason it was a mistake to make the escape wheel or the pallets of gold.

The best American watches had endstones of diamond throughout the escapement. Although a nice refinement, except with heavy balances, this is not really justifiable. Even then, if the ruby endstones are correctly cut with respect to the optical axis they will, in general, stand even the working conditions of a marine chronometer, where a large and heavy balance has small pivots with possibly a small radius of curvature at the pivot end.

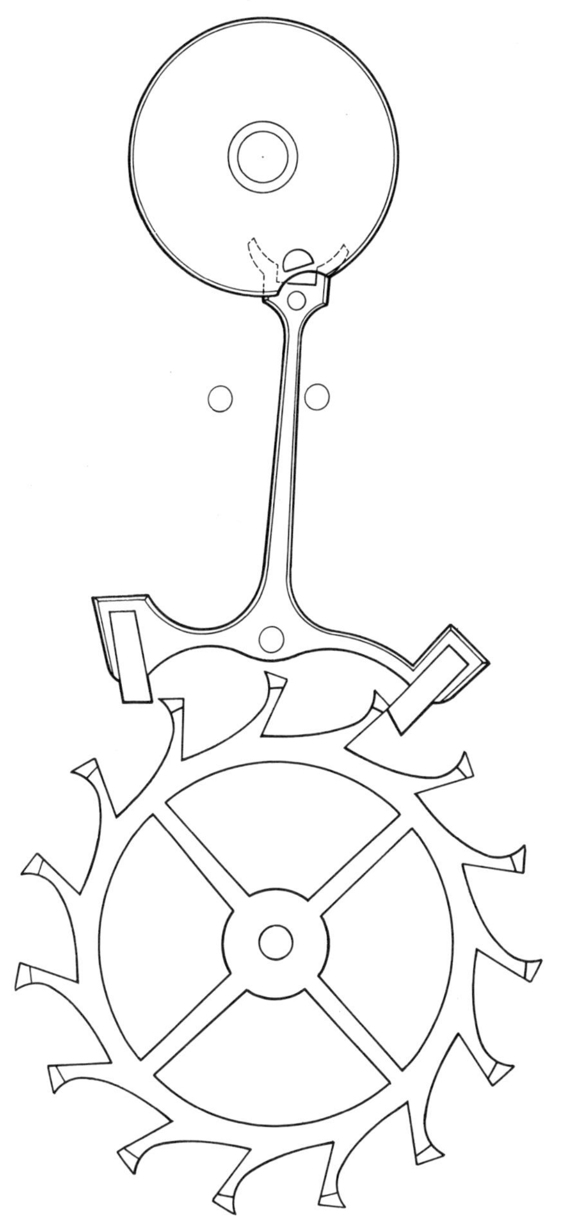

Fig. 26 Club tooth lever escapement. (Escape wheel revolves clockwise.)

10 *Keyless Winding*

One of the first attempts to overcome the need for a separate key to wind a watch was mentioned in 1752 by Pierre Caron, afterwards to become famous as Beaumarchais, the dramatist. His father, also a watchmaker, had made a watch for Madame de Pompadour which fitted into a finger ring. It was 9 mm in diameter and he wrote of it: 'To render this ring more commodious I have contrived instead of a key, a circle round the dial carrying a little projecting hook. By drawing this hook two thirds round the dial the ring is re-wound, and it goes for thirty hours.'

However, in 1820 Thomas Prest, who was John Arnold's foreman, took out Patent 4501 for 'a new and additional movement applied to a watch to enable it to be wound up by the pendant knob, without any detached key or winder'. The pendant button was mounted on a shaft, which also carried a pinion. This pinion, through a short train of gears, turned a wheel mounted on the barrel arbor. Unfortunately, this system was only applicable to going barrel watches and since the fusee was considered virtually essential by English makers the mechanism was not generally taken up. However, the idea was to be the pattern in the future.

Another early form of keyless winding was pump winding. The patent of the year 1793, taken out by Robert Leslie and numbered 1790, reads:

> 'A method of winding up a watch by the pendant. On the square where the key should go is a ratch; the pendant being alternately moved in and out, turns this ratch by means of two clicks on either end of a fork fastened to the pendant.'

Edward Massey took out another patent in 1824 and Viner also used a rack keyless mechanism with pump action. However, with all the foregoing types a key was still needed to set the hands, so that they were not truly keyless. The first watch that could be wound and set through the pendant appears to have been made by Louis Audemars of Le Brassus in 1838. He was followed by Adolphe Nicole in 1844, who took out a patent in England, No. 10348.

It was the work of Adrien Philippe (1815–94) which led to the watch that is both wound and set by means of the button, without the need to push or pull any other piece to set hands. He first offered his watches to the makers in Paris in 1842, but met with little encouragement. However, it was at the Exhibition of 1844, when his invention won him his first medal, that his work was appreciated by the Comte de Patek. A few months after the Comte offered him a partnership in his firm at Geneva, and thus was born one of the most illustrious watchmaking companies in the world, that of Patek Phillipe. This firm, one of the finest in the world, is still going strong with an unrivalled reputation. Two of their watches are shown in Plates 85 and 86.

11 *Guillaume and the Balance Spring*

It is strange that modern horology owes an enormous amount to Dr. Charles-Edouard Guillaume, a man who was not a horologist but a physicist. However, he came from a horological family and a horological district. He was born in Fleurier in the upper valley of the Swiss Jura Mountains, a region that also gave birth to Berthoud and Breguet.

After graduating from the Polytechnicum at Zurich and serving as an artillery officer, he joined the Bureau International des Poids et Mesures. Here he studied the mercury thermometer in great depth. In 1891 he began to research the characteristics of nickel from the point of its use as an alternative to the costly metals such as platinum and iridium used at that time as the standard lengths.

About 1895, the observations of the Director, J. R. Benoit, drew attention to certain alloys of iron and nickel, and after years of experiment Guillaume determined the constituents of an alloy which virtually remained constant in length despite the normal changes in temperature. Subsequently called Invar, this alloy was composed of 35·6 per cent nickel with iron. This has been used for pendulum rods ever since. Next, Guillaume turned his attention to balance-spring material and eventually developed Elinvar. This was another nickel–iron alloy, but with a considerable amount of chromium and a little tungsten and carbon, to obtain better mechanical properties, with 1 per cent or 2 per cent manganese added to assist in the working of the material. Used for balance springs it greatly eased the problems of temperature compensation, for it was possible to get results good enough for average usage without the need for a bimetallic balance.

A further development was Ni Span C, which together with Isoval and Nivarox, constitute the three types of balance-spring material in use today. All have fairly similar constituents and all are basically improvements of Elinvar, without departing to a major extent from the original. This is not to say that they are not vastly improved as compared with Elinvar, they are. Elinvar was soft and did not give a good balance action, all other things being equal.

Guillaume died in 1938, after being given many honours, one being the Nobel Prize in 1920, he also gained the gold medal of the British Horological Institute.

12 The English Watch Reaches its Peak

At the end of the 19th century the English watch still had an unrivalled position in the world. Great firms like Dent and Frodsham, Nicole Nielsen, Jurgensen, S. Smiths etc., turned out the finest of work (see Figs. 27a and 27b). Unfortunately, however, the English were satisfied with their way of working and did not feel the need to modernise their methods and employ an ever increasing number of machines, as did the Swiss and Americans. The result of this was that the industry became depressed and firm after firm went out of business. By the time the Second World War arrived, the industry was so depleted that it could not be successfully revived, even for the minimum requirements of the war effort. Lives had to be risked to obtain watches from Switzerland. After the war, an effort was made to revive the industry by setting up a National College of Horology. This produced about a hundred highly skilled technicians, designers and managers, before it was finally closed for lack of support from industry.

For a while S. Smith and Sons contrived to manufacture watches, but except for one or two of the less expensive variety, their efforts eventually ceased. Plate 80 shows the first prototype of a self-winding watch developed in Smith's own laboratories designed by P. W. Amis and made by hand by the author in collaboration with P. W. Amis.

Today and Tomorrow: A Revival of the English Watchmaking Tradition

Some individual craftsmen are beginning to make watches again. One such is a year watch made by Antony Randall, a young English

Fig. 27a Front view of triple complicated tourbillon, made by Nicole Nielsen for S. Smith and Sons. It is a split second chronograph, perpetual calendar, minute repeater.

Fig. 27b Movement of the triple complicated tourbillon. The tourbillon carriage is unusual in that it drives the chronograph work from the edge of the carriage.

watchmaker, trained in a Swiss school. Year watches are exceedingly rare, only one other example being extant. To make a watch that only needs to be wound once a year is a testing problem, solved only with the greatest difficulty.

Another superb craftsman is George Daniels. His latest masterpiece is a one-minute tourbillon, somewhat in the style of Breguet, with a constant force device in the train. He has also designed a type of Robin escapement with two escape wheels, giving impulse to the balance in both directions. This has all the advantages of the lever escapement and the chronometer rolled into one. It is in fact the last logical step in the evolution of escapements, the only thing that is surprising is that it had to wait until 1977 to be invented. The escapement has its roots in Breguet's 'escappement naturelle', not surprisingly, since Daniels is considered to be the world expert on Breguet.

13 Watchmaking in Other Countries

Swiss Watchmaking

The history of watchmaking in Switzerland begins in Geneva. How it started is not so certain, but is probably due in part to the influx of Huguenots, and also no doubt to the sorties of native Swiss into the established centres of watchmaking. By 1556 there are no less than fifty-two notary acts in Geneva relating to watchmakers during the 16th century, the oldest dating from 1556.

Apart from the magnificent enamelling done in Geneva, there is nothing in the early Swiss watches to make them outstanding. Nor did Switzerland have the outstanding figures, like those who blazed a trail in France and England. Technical innovation was never the strong point of the Swiss, and cannot be said to be so even today. The enormous talent of the Swiss has lain instead in, (1) their attention to detail, (2) their ability since the middle of the last century to produce watches cheaply by machinery, (3) their ability to produce these special purpose machines – to an accuracy unequalled elsewhere – and (4) their extreme specialisation.

Another feature that cannot be too strongly praised is their supremacy in complicated work, mainly situated in Valle de Joux. So successful were they in this field that it was common for ebauche to be sent from England to be fitted with complications, especially repeating work, thence to return to England for the rest of the watch to be finished. There was equally a flow of balance springs from England to Switzerland, until the latter years of the 19th century.

This is not to say that Switzerland did not have her own superb makers, she did. One of my favourite watches, shown in Plate 52, was almost certainly made by Houriet (Jacques Frédéric), 1743–1830, who was the founder of precision watchmaking in Neuchâtel. This is the magnificent tourbillon signed Hunt and Roskell. Abram-Louis Perrelet (1729–1826) is another famous name and his contribution to the invention of the self-winding watch is not disputed. Other famous names are Sylvain Mairet (1805-90), Louis Richard (1812–75), Ulysse Nardin (1823–76) and Frédéric-Louis Favre-Bulle (1770–1849).

Whether or not Breguet should be claimed by them is always a sore point with the Swiss. He was born in Switzerland and spent part of his working life there, when France was torn by the Revolution. However, he was trained in Paris and spent all his life there except for a period in England and the time in Switzerland already mentioned. Ferdinand Berthoud was also a Swiss. He was born in 1727 at Plancemont, over Couvert in the Canton of Neuchâtel and died in Paris in 1807. He left Switzerland for Paris when he was eighteen.

Jean-Moyse Pouzait was another famous Swiss maker. He invented the independent centre seconds watch in 1776 and also a lever escapement in 1786 that bears his name. There was also Henry-Louis Jaquet-Droz (1752–91), who was famous for his magnificent automata, among which were many watches with complications such as singing birds.

Antoine Tavan, although born in France in 1749, did most of his work in Geveva, having gone there once his apprenticeship was over. He is best remembered for his escapement models of which there were ten, three of the escapements being his own design. He was also a famous chronometer maker.

The Beginnings of Mechanical Manufacture in Switzerland

The beginning of the 19th century saw the beginnings of mechanical manufacture. About 1820 Humbert et Darier set up a factory for the production, by the use of machines, of rough movements and wheels and pinions. In 1854 Ritor, an ebauche maker of Geneva, conceived the idea of the die set, a two-pillar block in which the punch and die of a press tool were mounted. The use of a shaving tool, through which the relatively rough blank from the press tool was passed, originated it is thought from Ingold, a famous engineer. Ingold specialised in tooling for the watch industry and it is he who is credited with introducing watch-duplicating machinery in the U.S.A.

By 1875 the decorative chamfers on bridge and cock edges were also being put on by means of a press. The Swiss were further stimulated by the work being done in America in the way of mass-production. On attending the Exhibition of Philadelphia in America in 1877, the Swiss went directly back to Switzerland to introduce the new methods which they had seen. There was opposition to the new idea of 'mass-production', as there was in every industry in every country, with the exception of America, where skilled labour was in short supply. It is to the credit of the Swiss watch industry, (and the reason that they remained active while the industry in England dwindled to vanishing point) that they faced up to the facts and accepted the machine with all its ultimate benefits.

Within the space of half a century the industry was transformed. Ultimately, machines became so sophisticated and specialised that, despite the fact that the machine-tool industry was a separate entity with its own profits to make, they were controlled and not allowed to export key watchmaking machines.

Switzerland was particularly suited to the production of watches. The long cold winters meant that for six months of the year farmers could turn to another occupation. With few raw material resources, a labour intensive product was particularly suitable. It is probably also true that the national temperament is tailor-made for such work, for I think few would argue that the Swiss people are patient and painstaking.

In the 1930s all manufacturing countries were under threat. Many watch companies, as in every other sphere, contributed to their own downfall by embarking on a murderous price war. But for the action of some far-sighted men in Switzerland the whole watch industry could have foundered. These men co-operated and drew up a statute of sound trading. The Swiss Government approved and an autocratic organisation was set up. As they had over the matter of introducing machinery, the industry saw sense and were persuaded to sell only at economic prices. The prices of all the components that were made by specialists were negotiated a year ahead. Prices were related to the cost of production and there was no price cutting. It meant rigid control, but it saved the Swiss industry from ruin. With a monopoly of world markets this system was satisfactory, but when serious competition came from Japan and Russia, then cost became a factor outside Swiss control. In 1965, when the old statute ran out, competition was already so fierce that it was not renewed.

Things have now changed to the extent that Swiss firms will actually set up complete factories abroad, with ready designed and tooled calibres, so that the whole difficult process of producing watches can be started with the minimum of labour pains. Now, day after day, automatic lathes produce parts correct to a thousandth part of a millimetre. Parts not only go together without fitting, but are actually put together by machines. The skills have not disappeared, however, they have been shifted from one branch of the industry to another. The toolmaker of today regularly performs miracles. The making of checking and measuring equipment is an enormous industry in its own right. Press tools have become so sophisticated that components come from them better finished than they could be by hand. Diamond tools facet parts to a mirror finish, doing in a second what once took minutes. Hard plating produces durable finishes that last a lifetime.

It is probably now true to say that, apart from the attentions of a skilled adjuster, there is little to choose between the finest machine-made watch and one finished completely by hand fitting. In fact, if one allows for component selection it is possible that the machine-made watch could be better. Such is progress!

American Watches

Naturally enough, the first American watches were made by immigrant craftsmen from England, later joined by watchmakers from Switzerland and Holland. Early watchmakers, however, had a difficult time without the necessary support from specialists in their field. However, the first men in the field were not to be discouraged and a slow growth resulted in the setting up of the first watchmaking factory by Luther Goddard in 1809. Before he retired in 1817 no less than 500 watches had been turned out. This is a considerable number and probably means that movements in the 'rough' were obtained from England.

The first attempt to make a completely American watch was made by James and Henry Pitkin of Hartford, Connecticut, in 1838. Although this business foundered in 1841, this was not primarily due to technical difficulties, thus showing that an all-American watch could be a successful undertaking.

Only ten years were to pass before the setting up of the American Horologe Company. This was the result of the coming together of two brilliant men, Aaron Dennison and Edward Howard. Both men had ideas of producing watches by automatic machines which produced interchangeable components.

The first watches were produced by them in 1853 under the name of the Warren Manufacturing Company. The watches were full plate of the English style and sold for forty dollars. A year later the firm moved to Waltham, where under the name of The Boston Watch Company, some ninety men produced thirty watches a week.

However, the dream of producing all the parts in the factory caused endless difficulties and by 1856 Dennison and Howard split up. Dennison stayed as work's manager when a watch-case manufacturing firm bought the foundering company. In 1870 Dennison left the firm of Tracer Barker and Co., which eventually became the Waltham Watch Co. He came to England, where he founded the famous case-making company that still bears his name.

Howard, however, was not disposed to give up and by 1861 he had renamed his company the Howard Clock and Watch Co. In 1881 Howard retired, living on until 1904. He was the man who established the American precision watch.

The first watches produced by the Howard Watch Company were full plate and had uncompensated balances with flat springs. This was followed by a much superior design, as advanced as any of the time, which was three-quarter plate and had a patent barrel, designed to prevent damage to the gear train should the mainspring break. The barrel was stationary and the arbor was both rotated during winding and gave power to the train via the winding ratchet. To ensure that the watch did not stop during winding, maintaining power had to be employed. This was followed by a keyless watch, also employing a safety barrel and was jewelled in the English style.

Superb watches were to be made in America by Waltham, Hamilton and Elgin, comparing favourably with the best made anywhere and at any time. Waltham's finest watch was the 'Riverside Maximus'. It was fully jewelled, with diamond end stones fitted to the escapement and had beautifully damascened nickel silver plates. The train wheels were a low carat gold.

Elgin was to produce a deck watch fitted in a small chronometer-type case, complete with gimbals that were really superb (see Plate 76). It was free sprung with a guillaume balance and performed at least as well as many a marine chronometer, although probably not capable of keeping its rate for a comparable length of time.

Hamilton was to finally produce what is probably to be the finest portable mechanical timekeeper ever to be made. This was a marine chronometer, produced entirely by mass-production methods, which was developed and produced in an exceptionally short space of time. W. O. Bennett was the man who masterminded this operation, the same man who later backed the Bulova tuning-fork watch, 'The Accutron'.

Although the American watch cannot claim to have been better than either the English or the Swiss watch at any time, America certainly showed the way with respect to production techniques.

As already mentioned, at the Philadelphia Exhibition in 1876, the Swiss were greatly impressed and their methods of production were reorganised as a result. At the turn of the 19th century, production lines existed in America where raw material was fed in at one end and a finished plate came out of the other. Safety precautions were incorporated into the line, whereby if a tap or drill broke, an indication appeared and the line stopped.

Although American watches were made by the most modern of methods, the output of the Swiss was never equalled. The degree of specialisation achieved by the Swiss was never matched elsewhere.

At the heyday of output, for instance, just four factories produced almost all the escape wheels for the Swiss industry. The day came eventually, when the famous American names were put to watches imported from Switzerland and for a while it looked as if watch production in America would cease. The only reason that this did not finally occur was because of the electronic watch which favoured a return to American manufacture.

To add to the troubles of the American watch the Japanese watch industry, helped in its recovery after the war like the rest of the Japanese industry, began to present a strong threat to the watch producers of all other nations.

Watchmaking in Germany

The earliest known clockmakers' guild was founded in Annaberg in Saxony in 1543 but although the German watch industry was among the first, watches virtually ceased to be made during the Thirty Years War (1618–48) and the once vigorous industry lost out to the French.

What there was of the industry in the early 1800s was concentrated around Friedburg by Ausburg, Pforzheim, Silberberg in Schlesien, the Black Forest, Ruhla in the Turinger Wald and Glashütte, a mountain village thirty miles from Dresden. Until 1845 this village, with about a thousand inhabitants, existed mainly by farming and was very poor. However, a great man was to come forward who was to transform the lives of those who lived in Glashütte – Ferdinand Adolph Lange. Ferdinand Adolph Lange was a remarkable man. Born in Dresden on 18 February 1815 he went to Paris in 1840 where he worked for five years for the chronometer-maker Winnerl. Lange studied the watch industry in France and Switzerland and on his return to Sachen he began negotiations with the government to start a watch industry in the poverty stricken area of Glashütte. He was successful and in 1845 with a 30,000 Reichsmark loan, he started the production of watches with about twenty to thirty workers. These men were unused to such work and progress was slow. It took two years before the first watches were produced. Lange's first watches were pin pallet, first with steel pins, later jewelled.

These types were made during the period 1845–51. He also adopted the metric system of measurement. It was another thirty years before France and Switzerland followed suit.

However, during this period the characteristic Lange escapement was developed with its club-toothed, gold-coloured escape wheel. The enclosed pallets are also gold coloured and jewelled, with the impulse face of the entry pallet concave and the exit convex. Banking was provided by a pin on the underside of the pallet frame, near the entry pallet, working in a hole in the front plate. The ruby impulse pin was set directly into a strengthened part of the balance arm and the polished steel safety roller mounted on the staff. The two arm bimetallic balance had compensation screws and quarter screws. The latter were given the necessary friction in their tapped holes by a method unique to Glashütte watches. The rims were finely slit through the quarter screw holes, the slit reaching the hole on either side. This gave a slight spring, if the tapped hole was correctly matched to the thread on the quarter screws. These balances were made by Griesbach in Glashütte. The balance springs were steel. The fastening of the barrel ratchet wheel to the arbor, by means of an offset screw, is also characteristic.

The Lange factory flourished. Lange took training very seriously and designed and made the tools that made the watches. A case factory was also established in Glashütte to make gold and silver cases. Chronographs were produced from 1863 onwards and in 1866 he was granted an American patent for a repeating watch. Lange died in 1875 after thirty years of hard and successful work. His two sons, Richard and Emile, carried on his work. In 1885 Richard developed the first of the Company's self-winding watches, but it was not until 1892 that the first marine chronometer was made in the Lange factory. One of these marines, No. 795, took second place in the 1934 Hamburg trials.

After 1870 the higher grade Lange watches were made with removable barrels. The top plate was slotted and a bush used to locate the top end of the arbor. Removing two screws allowed the removal of the bush. In 1866 Lange patented a setting mechanism which allowed stem operation without disturbing the hands. Keyless wound watches are, however, later in date.

Lange watches are difficult to date, but one point to remember is that from the beginning of 1888 the German government required that gold cases be marked 14K or 18K on both the case and the bow. The Lange factory produced some exceptional watches. A number of tourbillons were made which are today snapped up at high prices in the unlikely event that one comes up for sale.

Fig. 28 English keyless watch with Glashütte escapement. (Recased by the owner.)

Watches were also produced with constant force escapements, others had three or four complications.

It is interesting that the Glashütte work was held in such high regard that some English makers even had their escapements fitted complete. See Fig. 28 and Plate 67.

The Russian, Japanese and Chinese Industries

The challenge which made the Swiss accept free enterprise again came from two countries, Russia and Japan.

Russian Watchmaking

When Russia overran Germany in 1945 she acquired both the men and the machines to boost her industry in a way that few thought possible. She decided from the start not to make pin pallet watches and by as long ago as 1968 she was producing 35 million watches a year. These were produced in five major factories, the biggest employed 6,000 people! Each of these factories was virtually self-sufficient and competed rather than co-operated with one another, and concentrated on a few models to satisfy the home market whilst supplying a surplus destined for the markets of the world. A large development and research organisation was built up. Eventually Russia even produced a marine chronometer, to the surprise of many.

Up until the 1930s Russia had no watchmaking industry. Protracted negotiations with the Swiss in 1935 to obtain their help to set up a factory led to nothing. The Russians had, however, already turned to America. The Amtorg Trading Corp. purchased practically all of the equipment in the Duber-Hampden Watch Works, Canton, Ohio in 1930. Mr. A. Vladimirsky of Moscow, Director of the Russian Watch and Clock Industry, superintended the dismantling of the plant and its removal to, and setting up in, the Russian capital. No less than twenty-four departmental heads of the Canton Plant accepted contracts from the Russian government and worked in Moscow for about a year, after which they were replaced by the men they had trained. Apparently sixteen ligne chromium-plated pocket watches with fifteen jewels were sold in the shops for the equivalent of twelve dollars.

The Rise of the Japanese Industry

Japan had no record of serious watchmaking until the close of the 19th century, although the Paperweight Clock (Plate 56) is really a watch. In the 1960s, however, watches of such quality began to be turned out that the Swiss markets began to be affected. Probably the most famous firm is that of Seiko, where every single part is made by the company. This firm was also the first to produce a quartz wrist watch in production quantities.

Chinese Watches

As if to underline her place as a major power in the world, China has also created a watch industry. Two wrist watches, both for men, have been created and represent a very creditable effort for a young industry. What her intentions are in the future is not known.

14 Special Watches

Cheap Watches

Until about 1750 owning a watch was a sign of wealth. Obviously a great market remained untapped, as long as this situation continued. Some cheap watches were being made in Geneva in the mid-18th century, but it was not until 1810 that serious attempts were made by Frederic Japy in France to produce watches by factory methods. These still required a fair amount of hand finishing, however. It is to Georges Frederick Roskopf (1813–89), that the credit is due for the first 'poor mans watch'. Roskopf was German born and was quite determined to succeed in his dream of a watch that could be made to suit a slender pocket. The first Roskopf watch, the name stuck to the type of watch until quite recently, was assembled in Chaux de Fonds in 1868. It was keyless wound but had no provision for keyless setting of the hands. These watches are now a collector's item. The example shown in Plate 62 is typical of the true Roskopf. It has distinctive hands and an enamel dial. The chief characteristics of the movement is that there is no centre wheel and that the escapement is pin pallet.

It is not surprising that the Americans were soon in the thick of the battle to produce and sell cheap watches. Amongst these were the 'Waterbury', probably the most ingenious cheap watch ever produced, the 'Auburndale Rotary', the 'Ingersoll' and the New York Standard Watch Co.

'The Waterbury' was designed by D. A. Buck in 1879 and the first sold for three dollars and fifty cents (about seventeen shillings and sixpence at the time). The early watches were 'long wind' with mainsprings twelve feet long and the movement revolved within the case, thus giving a 'tourbillon' effect. There were only fifty-seven parts in the watch, probably a record that has never been beaten. The watch broke all the rules and was obviously the work of a genius. The escapement employed was the duplex, an escapement considered then (and now) as too delicate to perform properly for long, the slightest wear resulting in difficulties. A 'Waterbury' wound today will usually start and perform without difficulty. Epicyclic gearing was employed and the layout is in fact difficult for most people to understand. The 'Waterbury' cost five shillings in 1900, the same as the 'Ingersoll' watch.

The 'Auburndale Rotary', produced in 1877, was also designed to have a tourbillon action. This was sold at ten dollars to the trade. Once again the whole movement rotated in the case. Unfortunately, a basic design fault caused the majority of these to be withdrawn by the firm and this no doubt contributed to the demise of the factory in 1883.

The New York Standard Watch Co. also tried to produce a cheap watch. This was characterised by a movement with a worm-driven escape wheel and with a curious lever escapement, with the escape wheel mounted like a verge escape wheel. This design was abandoned within the year (see Plate 66).

Robert Ingersoll sold his first dollar watch by mail order in 1892. By the time the American company failed in 1922 over 70 million had been sold, still at the same price that they were in 1892. The early Ingersolls had a dummy winding button but improvements were soon made.

The fascination of the tourbillon led the Swiss to produce the Mobilus. The design was patented in 1905 by J. Burtin. The revolving escapement was usually displayed at the dial centre or through the glazed centre of the back. The balance was not at the centre of revolution of the carriage and was small for the size of the watch. These watches now sell for prices that would astound their original makers! (See Plate 74.)

France and Germany also produced cheap watches, as would be expected. Japy Frères exhibited one at the Paris Exhibition in 1889, but it would appear that none have survived. Many watches were made by Junghans, but are mostly just marked foreign. Thiels produced an imitation 'Waterbury' in the 1890s but it needed winding every twelve hours, a return to the characteristic of the earliest of watches.

England also made an attempt to combat the imports of cheap watches by producing the 'John Bull', made by the Lancashire Watch Co. in 1909. They retailed at five shillings each, the trade price being three shillings and ninepence. The movement was marked 'British made by British Labour'. The undertaking was not a success, however, and the factory as a whole failed in 1911 after 5,000 had been sold.

The Automatic Watch

The automatic watch has a longer history than one might think. It is possible that there were earlier self-winding watches than those produced by Abraham Louis Perrelet of Le Locle in 1770, but there is no doubt as to his contribution. Breguet and Recordon were the first to purchase his watches and it is to these three men that the development of the self-winding pocket watch must be attributed. The winding device consisted of a weight, pivoted in the centre of the movement. The weight rotated through a full circle, with winding in both directions through the fusee. Breguet produced a number of self-winding watches after 1777, calling them 'Montres Perpetuelles'.

In 1780, Louis Recordon, Breguet's business associate in England, took out patent No. 1249 in London. Subsequently, the winding device became known as the pedometer wind. A number of makers were to produce such watches in the years that followed but nothing much happened until the idea was taken up again for a short time when A. Von Loehr, a Viennese engineer, took out his patent in 1878.

The self-winding wrist watch was what the world was really waiting for, once the wrist watch came into being in about 1910. It was John Harwood, an Englishman, who began experimenting in 1917 and eventually arrived at a self-winding wrist watch, the design of which suited mass-production techniques. His Swiss patent covering this was applied for in October 1923. The idea was to house a $10\frac{1}{2}$ ligne movement in a 13 ligne case. Hand setting was accomplished by turning the bezel, thus doing away with the necessity for a stem passing through the space around the movement needed for the housing of the rotating weight. The weight was pivoted in the centre of the movement and moved between stops. One could not wind the watch without shaking it since there was no manual wind provided (see Plate 79).

The 'Wig Wag' and the 'Rolls' watches

The patent for the 'Wig Wag' was taken out in 1931 and it was manufactured by Louis Muller and Cie SA Bienne. The movement was held by two arms which were pivoted in a framework fitted in the case. The movement had a possible lateral displacement of about 2 mm and as it moved, self-winding took place. The 'Wig Wag' watch was relatively short lived but another watch, the 'Rolls', built somewhat on the same lines was produced by Messrs. Leon Hatot SA, Paris. Here again, the movement slid freely in a carrier and was located by ball bearings. Self-winding was affected through levers and a ratchet. Hand setting was by a normal button, but could only be done once the outer case had been opened. The movement had a total displacement of 3 mm.

The Blancpain SA factory at Villeret in the Bernese Jura produced 6,000 of the 'Rolls' watches up until 1932 (besides 14,000 Harwoods).

The Rolex Perpetual Movement was the first really practical and long-lived automatic. Made in 1930 it had a semi-circular weight, pivoted at the centre of the movement. It moved through the full 360° but wound in one direction only. The reduction between the weight and the barrel was achieved by gearing and this eventually proved to be the favoured method.

The Bidynator was made in 1942 by Felsa of Grenchen, Switzerland. This watch represented the next phase of development, when the weight rotated the full 360° and also wound in both directions. This was achieved by the use of a rocking pinion, which is urged by the pinion mounted on the weight, into mesh with either of the two first gears in the self-winding train.

The Ultra movement represented the first of the last fundamental group of automatic watches. Instead of a gear there was a cam mounted at the weight centre. A long double spring, shaped like a W, reached across the block that carried the self-winding work. It was mounted on a steel plate and could slide backwards and forwards by a limited amount. The back and forward movement was caused by the cam moving two small jewelled rollers that were mounted on the same steel plate. The two extremities of the W spring were shaped into clicks and these alternately pushed and pulled a fine-toothed ratchet wheel. This had a pinion integral with it that drove the barrel ratchet wheel direct. Here the major reduction between the weight and the barrel was achieved through the geometry of the moving or sliding parts, as opposed to through the gearing.

Refining the Automatic Watch

Thereafter changes to self-winding work concentrated upon the following refinements.

1. Changing the geometry of the weight to achieve maximum winding torque with minimum mass.
2. The modifications of the slipping attachment in the barrel to give uniformity of torque output at the barrel. (Early attempts to put clutches on the barrel ratchet wheel were soon to become obsolete.)
3. Shock protection for the weight by design of the weight support, so as to make it flexible, by designing the weight so that it was flexible, or by mounting the weight on a ball race that could take the forces imposed without damage.
4. Easy disassembly from the main part of the movement.
5. Indication of the state of wind by inclusion of an up, or an up and down mechanism.

It is interesting that despite all the progress made and the near mechanical perfection of the automatic watch, the last problem comes back to lubrication. The more efficient the self-winding is, and it needs to be efficient for sedentary people, then the more the mainspring will slip when the watch is worn during active periods. This slipping of the mainspring can lead to serious wear in the barrel. There must be restraint or the mainspring will not be sufficiently wound and this restraint means friction and probable wear. As long as the lubricant is present, and is not contaminated with wear particles, all is well. Once it becomes contaminated a rapidly deteriorating set of conditions arise.

The last experiments along these lines include permanently sealed barrels. The spring is frequently dry lubricated and a heavy grease introduced between the spring and the barrel wall. In the event of failure the barrel is exchanged as a complete assembly.

The self-winding watch appeared to be in an extremely strong position at one time, so much so that it could have been assumed that eventually nearly all watches would have become automatic. However, the advent of the quartz watch has changed all this and appears likely to completely supersede this extremely interesting family of mechanical watches.

Complicated Watches

Watches not only tell the time of day. One of the first complications was not unnaturally striking work and watches that strike, both in passing (and later at will), have been made from the earliest days. The same comments apply to calendar work, where the phases of the moon were often displayed at the same time. In the past, of course, the question of whether or not any particular night would be moonlit was much more important than it is today. Watches also played tunes, gave sidereal time, the equation of time, the time in selected cities of the world, the night sky, the temperature, and had alarm work fitted. Others might have singing birds. Some special watches combined nearly all of these features, a fantastic tour de force by the most eminent of makers.

Thomas Mudge, already mentioned in connection with the lever watch, was a man who was very much at home with complicated work. Recently the British Museum obtained a perpetual calendar watch made by Mudge in 1764. It is probably safe to assume that this was the first successful perpetual calendar watch. A perpetual calendar watch corrects for the differing lengths of the months, including February 29 every four years, and was a feat only attempted in a few clocks prior to Mudge's triumph. Mudge was also the first to make successful minute repeaters. One was supplied to King Ferdinand V of Spain and was made to fit into the head of a walking cane. This watch was also a clock-watch. Another was sold at Christies in 1960 and was housed in a beautiful triple case, pierced and engraved to allow the sound of the striking to be clearly heard.

Mudge was to suffer a tragedy similar to that suffered by Beethoven, for at a relatively early age his vision began to fail. However, he was such a clever and determined craftsman that one can detect virtually no difference in his work as his sight progressively worsened.

Breguet was also a master of complicated work, as can be seen in the watch he produced that was intended for Marie Antoinette. This was ordered in 1783 by an Officer of the Queen's Guard and was to contain every complication known at that time. No limit on the time taken to manufacture it or on its cost was imposed and all parts normally of brass were to be of gold. It was not finished until 1820 and by that time had cost 16,864 francs. Some idea of what this means is given by the fact that the gifted pupil, Michael Weber, made most of the mechanism for which he was paid ten francs an hour for 725 hours work. Thus it could represent 1,686 hours work, which would these days represent about eight years work. Such a highly skilled craftsman would expect today to earn about £8,000–£10,000 a year, so that the total cost today would be £64,000–£80,000. When one considers that currently the Audemars Piguet triple complicated watch costs £22,500 this sum is not surprising. Breguet's watch contained the following complications:

Perpetuelle winding
Perpetual Calendar
Equation of time
Thermometer
Centre seconds marking whole seconds

Indicator for state of wind
Minute repeater

The plates, bridges and all the train wheels were of polished pink gold. The dial, back and front covers were of rock crystal.

Breguet did not sell the watch. It remained in the family until, in 1887, it was sold by the widow of Louis Clement Breguet to Sir Spencer Bruton for £600. It is now on view to the public in the L.A. Mayer Memorial Foundation, Jerusalem, Israel.

A Patek Phillipe watch made in 1932 had a dial back and front and gave the following indications:

Mean time
Chronograph with split seconds
Up and down work for both barrels (going and striking)
Perpetual calendar
Phases of moon
Sidereal time
Equation of time
Sunrise and sunset
Night sky

In addition it was grand sonnerie, minute repeating, carillon on four gongs. The watch contained 110 wheels, 50 bars, 430 screws, 90 springs, 120 pieces of mechanism, 70 jewels, 19 hands and discs. It took five years to make, excluding the time taken to design it.

How such a watch was designed at all is something of a mystery. The parts for each complication cannot occupy their own separate space in the watch, but must be interconnected and although, in so far as possible, occupy separate layers, must encroach and interleave in very many instances. It is highly probably that many of the detail difficulties can only be worked out as the watch is being made.

A Dent watch, which may well be one of the most complicated English watches made, recently came up for sale in Switzerland. It is No. 32573 made in 1904. It has a gold case, a gold and silver dial one side and an enamel dial the other. It has the following complications:

Clock watch with grand and petit sonnerie
Minute repeater
Perpetual calendar on one dial

On the other dial

The signs of the Zodiac
Equation of time
The morning and evening stars
Moon dial
Sunrise and sunset
Moonrise and moonset

An ebauche in the British Museum suggests that these English watches went in the rough to have the complicated work done in the Vallé de Joux, in Switzerland, and on being returned were finished in the English workshops.

Even today a triple complicated watch is being made by a firm of Audemars Piguet. This is a minute repeater, has chronograph work and is a perpetual calendar. They are made at the rate of one or two a year and at the moment sell for £22,500. There is a waiting list of five years production.

Complicated Wrist Watches

In the late 1920s a small Swiss minute repeater, possibly the most complicated watch of its size ever made (the size is 11 ligne), was custom-built for an American, one James Schulz of New York in the late 1920s. It was made in a small Swiss home workshop in the French part of Switzerland just outside Geneva. Three Swiss watchmakers worked on it for three years. The watch is perpetual calendar, has phases of the moon, is a minute repeater and has chronograph work with a thirty minute register. The tonneau-shaped case is of platinum. When last recorded this watch belonged to a Robert B. McConnell who acquired it in 1964 and who prizes it above all others in his collection. He believes it to be the world's most complicated watch for its size and I am inclined to agree with him. Plate 78 shows a modern chronograph with phases of the moon and date work.

An outstanding achievement in the way of a modern complicated wrist watch is a self-winding date watch with chronograph work. This watch has been designed so cleverly that it is possible to lift off the chronograph work complete, this being carried by a sub-plate that can be removed by loosening just three screws. (See Plates 84a and 84b.)

Libertine Watches

For the first time a book is being produced on these watches, for sale on the open market. This underlines the change in attitudes today. In 1954 there was a sale of a number of these libertine watches. Nowadays catalogues show the scenes in question, but in 1954 the harmless aspects of the watch were shown and guarded statements made such as 'concealed automata', 'animated scenes', or best of all 'with interesting automata scenes'. And this was a mere twenty-four years ago! Only one of these watches, although there were about thirteen, bore an English maker's name – Johnson of London.

Such watches were mainly made by the Swiss and were of two types. In one, erotic scenes are disclosed when a cover is raised. The second type had automata, driven by repeating work. Amazingly enough there is now a modern development of the libertine watch. The crudest of statements appears on the dial of a wrist watch at intervals of fifteen seconds. The exhortation is inscribed on a nearly nude woman. One is assured that the watch is dustproof to ensure accuracy!

Small and Flat Watches

Very small watches are not a new thing. In the Chaux de Fonds Museum in Switzerland is a pre-balance spring, anonymous verge watch with gut fusee of 12 mm in diameter! Yet more astonishing is an even smaller watch, once in the celebrated collection of Sammlung Marfels. Again anonymous, its date is about 1650 and it is an astonishing 9 mm in diameter. Both dial and case were enamelled on gold. The whereabouts of this watch is not now known.

Much later, John Arnold produced a ring watch for George III, which was additionally a quarter repeater. It had a cylinder escapement and the cylinder was made of ruby. George III paid Arnold £500 for his trouble – which must have been considerable. Arnold was subsequently asked by Tzar Alexander to make another of these. Patriotically he refused, saying that he wished King George to be the only man to own such a watch.

Very flat watches are still being made by the firm of Golay in Switzerland. Some are worn as wrist watches, but others are set into gold coins. When finished, no sign is visible of the presence of the watch within the coin, except to those with a very keen eye, who know what they are looking for and can just distinguish the outline of the catch in the milled edge of the coin. When this is pressed one face of the coin flips up, as in a hunter watch, revealing the watch, complete within an inner case.

A revolutionary new movement has been produced by a Swiss firm, Bouchet-Lassale. The basic movement is only 1·20 mm thick and the self-winding version *with the rotor at the centre* is only 2·00 mm thick! The revolutionary feature is that all the moving parts are supported on one side only with the exception of the escapement. The barrel is edge supported by rollers and the other wheels by ball bearings. The claimed advantages of this design are as follows:

1. No alignment problems between bridge and plate
2. Fewer frictional losses
3. Simpler lubrication
4. Simpler assembly

The diameter is 20·40 mm (9 ligne). The movement has a fast beat rate – 28,000 per hour with a fifty-hour reserve.

Coin Watches with Skeleton Movements

Not only are ordinary movements, if one can call them so, fitted into coin watches, but so are skeletonised movements. It is as if to say, now that we have done the improbable let's do the impossible.

The plates, bridges etc. are frequently made of gold. This is also true of an automatic watch made by Patek Phillipe. The self-winding weight is 22 ct. gold, so as to give the maximum winding power with minimum size. Such watches are nothing less than miniature works of art.

The Electric Watch

The electric watch first appeared on the scene in a satisfactory form in 1957, made by Hamilton of America (see Plate 81). It is fitting that the firm who managed to produce 10,000 chronometers in three years without previous experience (and yet of the highest quality ever), should be the company to first succeed with the electric watch. Of course, the pundits prophesied that an electric watch was just a nine-days wonder, but thinking people realised that it was but the start of the long road to the death of the mechanical watch as an everyday item. Admittedly, the Hamilton electric watch had a delicate contact system and no watchmaker could really undertake its repair without guidance, but nevertheless it was a tour de force. However far seeing anyone was at that stage it would have needed second sight to realise that within twenty years the quartz wrist watch would be commonplace, sold even in the supermarket. Such is the present speed of progress.

When the electric watch first arrived, it posed no real threat to the mechanical watch. Its timekeeping could be shown to be marginally better, but against this it was more difficult to get serviced and needed a new battery every year.

The arrival of the first truly electronic watch, the Bulova 'Accutron', changed this and the mechanical watch was in jeopardy. However, the advent of the quartz watch finally put an end to the hopes of those who made mechanical watches since a good quartz watch can be guaranteed to keep time to a minute a year, a feat no mechanical watch can rival.

The 'Accutron' was the first watch to move away from the balance and spring as the controlling device. Instead it employs a tuning fork. This is made of Ni Span C and vibrates at 360 cycles a second. It drives a ratchet wheel directly. This has 300 teeth which are so fine that a 30× magnification is required to see them with any degree of clarity. At this speed of indexing many feared that either the indexing jewel or the ratchet teeth would wear rapidly, but in fact nothing of the kind has happened. This is to say the least surprising, until one realises the tiny forces involved. Even so 30,000,000 indexes a year is a daunting figure. The author's 'Accutron' has now completed 240,000,000 indexes! (See Plate 82.)

After the 'Accutron', nothing was ever quite the same again and before long the Swiss produced the first production quartz pocket watch. Not long after this the Swiss were pipped at the post by the Japanese, who were the first to put a quartz wrist watch on to the market.

From then on, efforts were mostly in the direction of making quartz watches less expensive. In the course of a few years prices fell from £900 to £30. Currently quartz watches of the LED type (light emitting diode) are on sale in some shops at £14. The market is in a turmoil at the moment with unreliable and reliable models indistinguishable by the public. Wild claims are made for watches fitted with cheap crystals and inadequate circuitry incapable of the timekeeping ability credited to them. However, this stage will pass. There is little doubt that reliability is really the prime consideration, since few people actually need the sort of timekeeping that the best quartz watches can provide. Ebauches SA have produced a very thin quartz watch that they call the 'Flatline'. This watch is 3·70 mm thick, thus combating the arguments of those who say that for flat watches the mechanical watch will never be replaced. Omega have also produced a 6 ligne ladies quartz watch which was shown at the Basle Fair in 1977. The battery is only 6·00 mm in diameter. Thus, all requirements are now being covered and still we are only at the beginning of the quartz watch story.

The day is probably coming when most quartz watches will be purchased set to time, to be thrown away before the need is ever felt to correct them. The watch repairer as he is known today will disappear to be replaced by those capable of restoration work on antique watches, which description will cover the mechanical watch and the electric, as opposed to the electronic watch. Quartz watches are also being produced with many complications, perpetual calendar, chronograph and some are, in addition, miniature calculating machines. It is, of course, impossible to predict what the future will bring. Perhaps a watch that is set into a tooth that will give the time audibly when desired – who dares to laugh!

Glossary

Adjusted Refers to work done on the balance and spring towards eliminating timekeeping errors due to changes that occur in position, temperature or the state of wind. Unadjusted does not mean to say that nothing has been done about these problems at the design stage and in the choice of materials, or that careful poising has not been done. What it does mean is that no further work, other than bringing to time, has been done in view of the individual behaviour because of the changes mentioned.

Adjustable Potence see Potence.

Affix A small bimetallic strip fitted to the main balance which may or may not be bimetallic, so as to affect small changes in the moment of inertia of the balance. This may be to overcome middle temperature error or to make the small corrections required, due to the residual errors of a 'self-compensating' balance spring and balance combination.

Airy's bar A fine adjustment arrangement for bimetallic balances invented by Airy. Two weights lightly sprung against the inside of the balance rim can be slid around the rims, so that they add to the amount of compensation as they approach the free end of the cut rim. A very useful adjunct which found favour among chronometer makers.

All or nothing piece A piece in a repeating watch designed to prevent incorrect striking. If the slide is not pulled all the way round or the push piece fully depressed, so as to ensure that the full number of blows are struck, the watch will not strike at all.

Amplitude The arc of swing measured from zero position to extreme position. Thus, a pendulum which has a total movement of 3° has an amplitude of $1\frac{1}{2}$°.

Anti-friction wheels Wheels arranged so that their outside diameters are the bearing surfaces of the pivots of balances or of wheels in the gear train. Thus, the frictional restraint is reduced by a factor given by the ratio of the outside diameter of the anti-friction wheel to its pivot diameters.

Anti-magnetic Having the property of resisting the effects of an applied magnetic field both during and after the application of the field – the latter being the more important. Without definition the term is valueless.

Apparent solar time The time as given by the sun, i.e. as would be shown on a sun dial. To convert this to Mean Solar Time one needs to know the Equation of Time.

Appliqué Applied ornament to a case, or chapters, numerals etc. to a dial.

Arbor A shaft with bearings. If integral with a gear it is called a pinion.

Arc The arc of a balance is twice the amplitude often denoted as 'turns' by watchmakers. A balance 'doing one turn' has an arc of 360° and an amplitude of 180°.

Assortiment A complete set of escapement parts.

Atelier Workman, especially in horology or allied trades.

Attachment The position that the inner termination of the balance spring bears to the watch.

Automaton watch A watch with pieces that move, usually at will when causing the watch to repeat.

Auxiliary compensation The extra compensatory device fitted to a compensation balance to help to eliminate middle temperature error.

Backslope On an arbor, the part that slopes from the shoulder to the next larger diameter. On a staff, the tapered portion below the balance seating and also the slope behind the pivot blending radius.

Balance Used to control the rate of revolution of the gear train. This is a wheel with one, two, three or four spokes known as arms. Used without a spring in early timepieces, it was not until fitted with a spring that good timekeeping became possible.

There are three main types of balance:

1. Plain, where the rim may or may not be furnished with screws.

2. Cut, where the rim is bimetallic and is cut so that there are two, three or four circumferential strips.

3. Ovalising, where the material is chosen for its unequal expansion in two directions at right angles to one another or is made of two different metals when either the arm or the rim is usually made of Invar.

Moving screws and or weights enables the degree of compensation to be changed, i.e. by moving them nearer to or further away from the point at which the rim is anchored.

Balance cock The cock that supports the balance staff.

Balance screws The screws that are provided around the rim of the balance and that are used to poise the assembly, to bring it to time, to affect the amount of temperature compensation or to enhance the appearance; all or any combination of these.

Balance spring The spring that in conjunction with the balance provides an oscillating assembly. It may be flat spiral, helical, have an overcoil, be a tapered helical, spherical, or be a combination of certain of these, and made of various materials.

Balance-spring stud The stud is the fixing for the outer end of the spring, or the top end in the case of a helical or spherical spring (that is the end that is fixed to the balance cock or bridge). The fixing to the balance is known as the collet. This is normally mounted friction tight on the balance staff.

Balance staff The arbor on which the balance is mounted is called its staff.

Banking The part against which the lever in a lever escapement, or the detent in the detent escapement, rests once escapement action is over.

Banking pin A pin used to provide banking.

Barrel The barrel that contains the mainspring.

Barrel, hanging A barrel that is supported at one end only, namely the top.

Barrel, resting A barrel that is supported at one end only, namely the bottom.

Barrows regulator See Regulation.

Bascule escapement A loose term for the pivoted detent escapement.

Basse-taille enamel Translucent enamel on a surface that has been engraved.

Bassine A smooth rounded-edge case.

Beat The escapement sound or 'tick'.

Beetle hand The type of hour hand that vaguely resembles the shape of the stag beetle. Used with a 'poker' minute hand.

Bezel The rim holding the glass.

Bimetallic Formed of two metals, in horology usually brass and steel, but not invariably so.

Black polish A polish so perfect that at certain angles it looks jet black.

Bottom plate See Top plate.

Bouchon A brass or nickel bush used as a bearing for pivots.

Bow The loop at the top of the watch by means of which it can be lifted or attached to the albert, chain, chatelaine or what have you.

Breguet spring A flat spiral spring with the last turn raised from the body of the spring and formed into a terminal curve, capable of manipulation with the object of reducing positional and isochronal errors.

Breguet hands The moon hands much used by Breguet. These hands tended to be more delicate than those used before Breguet introduced a new look to the watches of the time.

Breguet key See Tipsy key.

Brevet The French equivalent of Patented.

Bulls-eye glass A domed glass with a flat ground in the middle.

Bush See Bouchon. Also the act of bushing a piece to correct a bad hole or to lengthen the bearing surface.

Button, winding The serated cylindrical or spherical piece that is rotated by the fingers to wind the watch or set hands.

Cadrature The work beneath the dial.

Calibre The type of watch often including the size. It is really the brief specification of the movement.

Cam A piece which when rotated is so shaped that it can move another piece in a controlled manner. The shape may be on the edge of the cam or on its face.

Cannister case A drum-shaped case, similar to the tambour case but not hinged.

Cannon pinion The part of the motion work mounted on the centre arbor and able to move relative to it during hand setting. It meshes with the minute wheel in normal layouts.

Cap jewel An endstone.

Capped movement A movement with a dust cap.

Carriage The piece in a tourbillon that rotates and contains the escapement.

Cartouche dial A dial that has cartouches. Cartouches are shield-shaped pieces usually of enamel, which are applied to the dial but may be produced on the dial by engraving and different finishing (as on a matt dial when the maker's name is on a polished oval).

Centre wheel and pinion The wheel and pinion at the centre of the watch; it usually rotates once an hour.

Centre seconds A watch with the seconds hand at its centre – also in modern watches called sweep seconds.

Chaff cutter See Ormskirk and Debaufre escapements.

Chaise watch A very large watch, also called a travelling watch.

Champlevé A hollowed-out area of metal that can be filled with enamel or niello.

Chapters The hour numerals or marks on a dial.

Chapter ring The ring on which the chapters are put, and also the minute and possibly an hour ring with quarter-hour divisions.

Chasing Engraving in relief.

Châtelaine A chain for suspending a watch or a piece of jewellery. As well as the watch, the winding key, seals and trinkets were also often attached. Very often the decoration on the châtelaine was made to match the watch.

Chinese duplex A form of duplex escapement invented by C. E. Jacot in 1830. Commonly used in Fleurier watches intended for China. The teeth on the locking wheel are doubled so as to resemble a fork. Each of the two prongs is held in turn by the locking roller, so that impulse is given on each alternate full swing of the balance. Thus, if the train is 14,400 beats per hour the seconds hand will move forward once a second. Thus, the watch appears to beat seconds, although there is a slight movement of the hand each half second as the locking changes from one fork of the tooth to the next. There was even one version of the escapement that had three prongs on each tooth!

Chronograph A watch with a centre seconds hand that can be started, stopped and returned to zero at will. Also usually provided with a minute counter and sometimes an hour counter to show how long the centre seconds hand has been in operation. All these hands are returned to zero together. See also Split seconds.

Chronometer In England a chronometer is understood to be a watch or marine timepiece that has a detent escapement (although the word was used before this escapement came into use). Unfortunately the word has been misused by the French and Swiss to mean any watch that has obtained an official rating certificate. This has resulted in great confusion in the public mind.

Collet A ring-shaped piece. In watches it may belong to a hand, a balance spring or to a wheel.

Compensated Taken to mean temperature compensated. The balance and balance spring are so constructed that over a certain range of temperature errors are reduced, these errors being due to dimensional changes and to stiffness changes in the spring.

Compensated balance Usually taken to be a balance that has to provide all the compensation required, but could be a balance that only has to provide the residual compensation for the so-called self-compensating balance spring.

Compensation curb A bimetallic strip that carries the curb pins, thus moving them in some manner that is meant to compensate when temperature changes occur.

Compensation screws Screws that can be moved along the balance so as to distribute them in the manner that, in conjunction with the movements of the balance rim, will provide proper compensation.

Compensation weights Weights that provide the same facility as the compensation screws.

Complicated work Usually taken to mean additions to a timepiece other than striking work, e.g. calendar work, phases of the moon, equation of time etc.

Cone pivot A pivot shaped like a cone. Nowadays used in cheaper watches. Although known as conical, the acting end of the pivot ideally has a carefully controlled radius.

Conical pivot The name given to the normal balance pivot, which is in fact not conical at all, but has a parallel portion with a radiussed end and is blended into the next diameter of the staff with a curve.

Constant force escapement A special escapement with subsidiary spring providing motive power. This is wound at frequent intervals, thus the force at the escapement is virtually independent of fluctuations due to the imperfections of the gear train and the variations in the main motive force.

Consular case Understood to be the double-bottom case fitted with a high rounded glass, although the term is used somewhat loosely.

Contrate wheel A wheel to transmit motion from one arbor to another that stands at right angles to it; the teeth instead of projecting from the periphery stand out from the face of the wheel.

Conversion An escapement substituted for the original. Generally a lever escapement supplants a verge, a chronometer or cylinder escapement. Nowadays important pieces are being re-converted by replacing converted escapements with one closely similar to the original.

Coqueret A steel end plate mainly found in French watches after 1735. Screwed to the balance cock, it provides the end bearing for the top pivot of the balance. By 1770 it was in general use but was then gradually supplanted by a jewel bearing as was the English practice.

Count wheel See Locking plate striking.

Coupe-perdu escapement One in which unlocking of the escape wheel does not always lead to impulse. The Chinese duplex is such an escapement. Usually this is so that the seconds hand can move forward in an interval longer than that given by a normal escapement, usually one second with a Chinese duplex.

Counterpoised pallets Those that have an appendage specifically to help to poise the pallets.

Crank lever escapement A form of lever escapement in which impulse is conveyed to the balance through what is really a single-gear tooth. The notch in the end of the lever is in the form of a gap between two gear teeth, and the projection on the roller is made as a single pinion leaf. Invented by Massey of Liverpool.

Crank roller See above.

Crescent or passing hollow. The clearance cut into the roller of a lever escapement for the passage of the dart or guard pin.

Crown wheel escapement Another name for the verge escapement.

Curb pins or index pins. The pins that embrace the balance spring and when moved along, affect regulation.

Cut balance See Compensation balance, although not all compensation balances are cut. Some modern balances merely have affixes or become oval to affect compensation.

Cuvette An inside cover to protect the movement, hinged and sprung, often of brass even in good quality gold cases. Used on continental watches and often engraved with maker's name, directions of winding, hand setting, number of jewels, escapement etc.

Cylinder escapement or horizontal escapement. Perfected by Graham in about 1725, the balance is mounted on a hollow cylinder large enough in the bore to admit a tooth of the escape wheel. Nearly half of the cylinder is cut away where the teeth enter, and impulse is given by the wedge shape of the teeth as they enter and leave. The escapement is dead beat, frictional rest. The pieces inserted in the hollow cylinder on which the pivots are formed are known as plugs.

Dart Another name for the guard pin, from its shape on high quality hand finished English and Swiss watches; they were often screwed in position. See Guard pin.

Daily Rate The amount a watch varies each day from correct time.

Damascene Ornamental finishing of watch parts, especially on keyless wheels and the barrel bridge. This was a speciality of the American watch industry.

Dead beat escapement First used successfully by George Graham in 1715 but possibly preceeded by Tompion. Its feature is the elimination of recoil in the escapement by making the locking faces of the pallets circumferential with the pallet pivots. Although applied to the Graham clock escapement it is also a generic term for escapements without recoil.

Dead beat verge An escapement belonging to the Debaufre family, where two bevelled edge pallets receive impulse from a verge-type escape wheel.

Deaf piece A push piece in a repeater that when pressed prevents the watch from striking by taking the hammer blow. By this means the blows can be felt instead of heard. Not just for the deaf, since there are occasions when the noise could be a nuisance.

Debaufre escapement Invented by Peter Debaufre in 1704. Two ratchet tooth escape wheels are on the same axis, the teeth being staggered in relation to those on the other wheel. The balance staff passes between the two wheels and a pallet is fixed to the staff.

Deck watch A precision watch for use on a ship when establishing its position. It is normally large, say 18–24 lignes, and is used to transfer the time from the ship's chronometer or other time standard to the position where sightings are being made on deck. Also used instead of a chronometer where the vibration would otherwise be too much for the normal chronometer.

Depth The degree of intersection of two parts that work together, such as gear wheels or escapement parts.

Detached escapement An escapement which is not in contact with the balance except when unlocking, impulse or run to banking is occurring. Thus, the lever and chronometer escapement are detached whilst the duplex, cylinder and verge are not.

Detent A loose term for a stop generally confined nowadays to the detent of a chronometer escapement. Also, however, the fusee detent and the warning detent in striking.

Detent escapement Another name for the chronometer escapement, pivoted or spring detent.

Dial plate The plate to which the dial is fixed. May also be the bottom plate or pillar plate.

Differential dial Here the centre of the dial is a disc marked 1 to 12 and makes $\frac{11}{12}$ of a revolution per hour. An ordinary minute hand revolves concentrically with this disc once an hour and is thus always passing over the current hour.

Differential up and down work Differential gearing is required for one form of up and down work on going barrel watches.

Discharging pallet The pallet that affects unlocking in the chronometer escapement.

Divided lift In an English ratchet-tooth escapement, all the lift, or impulse, as on the pallet. In Sylvain Mairet's lever escapement all the lift was given by the teeth of the escape wheel. In the divided lift escapement, lift is shared so that some is given by the wheel teeth and some by the pallets. This is the family to which the club-tooth lever escapement belongs.

Dome The inside cover of a watch case which is like a second back.

Double bottom case The form of watch case most used to house full plate movements. The inner bottom or back is in one piece with the case band, and to inspect the movement this has to be swung out from the front of the case on its hinge, or as it should be called, 'joint'. Before this can be done the bezel must be opened and the bolt released.

Double roller The type of roller used in the modern lever escapement; it has one table that carries the ruby pin and another that provides part of the safety action arrangement.

Draw The characteristic of an escapement that keeps the lever or detent safely in the rest position once escapement action is over, despite all but the most severe jolts and jars. The part is impelled against the stop or banking by utilising the force that is needed to recoil the escapement. This is obtained by having suitable geometry at the pallet, so that in moving away from the stop or banking, as it is called, the escape wheel is forced to move slightly backwards. Although the knowledge of the feature draw was known, early makers, including Breguet, were reluctant to use it in the lever escapement since it seemed to go against what they believed in; that recoil escapements were bad and equally so was any recoil in any other escapement. John Leroux, in 1785, was the first to introduce draw in the lever escapement, but it was in fact present from the start in the chronometer, merely by the accident of the features of Arnold's particular design.

Drop The necessary small free movement of the escape wheel as it moves from one pallet to the other. Without this the escapement geometry would have to be perfect for the escapement to act without jamming, a manifestly impossible thing. However, the higher the quality – and by inference the more perfect the geometry – the less drop there can be and its amount is in fact a design decision. In a good-quality lever escapement drop is usually $\frac{1}{2}^{\circ}$ which at a diameter, say of 5 mm, means a linear movement of about $\frac{2}{100}$ mm. Drop is also the amount the chronometer escape wheel moves before it catches up the impulse pallet on the roller.

In a cylinder escapement, drop is the amount the wheel tooth moves forward on to the outside or the inside the cylinder after impulse is completed.

Dumb repeater A repeater without bells or gongs where the hammers strike a block mounted in the case.

Duplex escapement Commonly an escapement whose wheel has two sets of teeth. One long and pointed set provides the locking function by working with a small roller with a passing notch that is mounted on the balance staff. The second set of teeth stands up from the rim of the wheel and gives impulse to a pallet (which is sometimes jewelled) that is mounted on the balance staff.

Less commonly, two separate wheels are used to perform the two functions. Like the chronometer, the impulse teeth do not need to be oiled but oil is, however, required at the locking jewel. Although a frictional-rest type this escapement can, if not worn, perform extremely well.

Dust cap A cap to help to exclude dust fitted mainly to the movements of watches that are hinged to the case. They came into use in about 1715. Usually made of brass, though sometimes of silver. As a rule found in English watches.

Ebauche The movement blank or rough movement. In the early 19th century they consisted of the plates, pillars, cocks and bars with barrel, fusee, ratchet, ratchet wheel and assembly screws.

End shake Axial play. The clearance between the pivot shoulders and their bearing surfaces.

Endstone A thrust bearing made of jewel, against which a pivot end bears.

Endless screw See Tangent screw.

Engine turning A regular pattern machined with a special-purpose lathe. Even when the pattern consists of straight lines it is still called engine turning. Often covered in clear or translucent enamel when it is called Guilloché.

English lever escapement The pointed-tooth lever escapement where all the lift is on the pallets.

Entry pallet The pallet in the lever escapement where the escape-wheel teeth enter. If one looks down on an escapement and the wheel is rotating clockwise, then the entry pallet is on the left.

Equation of time The difference between solar time and mean time. Only at four times of the year do mean time and solar time agree.

Escapement The element in a timepiece that transforms the rotary motion of the gear train into the oscillatory motion required by the timekeeping element.

Escape wheel and pinion The wheel that gives impulse to the oscillating member, the balance, foliot, or balance and spring, either directly or through an intermediate.

Figure plate Another name for the Index Dial used with a Tompion regulator.

Finishing The work required to bring a watch from the ebauche state to the finished condition.

Fire gilding An amalgam of gold and mercury is used to coat the article to be gilded and the mercury driven off by heat to leave a strongly adhering film of gold.

Five minute repeater A repeater that strikes the hours and then a number of double blows for the number of five minutes past the hour.

Flags The pallets of a verge. See Verge escapement.

Flirt A lever or some device that causes a sudden movement to occur. It is usually raised slowly by a cam to be suddenly released, or may be under manual control.

Floating hour dial See Wandering hour.

Fly A governing device that consists of a vane fixed to a pinion, which in sweeping through the air slows the rate of rotation of the train of which it is part. Used in striking, repeating work, and in remontoires and constant-force escapements.

Fly back hand A hand that is caused to, or automatically flies back to, a zero position or to coincide with another hand.

Fob chain A short chain with a swivel or bolt ring to fix it to the watch and usually with a bar at the other end to pass through the button-hole. It hangs outside the clothing.

Foliot The earliest form of controller. It is bar shaped, and in watches usually has fixed weights at its ends. Its purpose is to increase the moment of inertia of the verge to slow its operation.

Forgeries The whole question of forgeries and fakes is excessively difficult, and becoming more so. Even top experts have trouble in this field. Many old forgeries are collected in their own right. Some, both old and new, are virtually impossible to detect.

Fork The fork-shaped end of the lever in the lever escapement in which is the notch.

Fourth wheel and pinion The fourth wheel is the wheel that normally drives the escape pinion. It often carries a seconds hand. It is a loosely used term. In an eight-day watch it may in fact be the fifth wheel, but would never be so called.

Form watch A watch whose case is made into a shape recognisable as representing something other than a watch case. See Index.

Four coloured gold Gold can be produced in a great number of colours according to its alloying constituents. Dials were frequently decorated with different coloured golds, sometimes as many as four. Cases were also treated in this way but less frequently.

Frame The main parts of a watch that support and locate all the other parts.

Free-sprung Without an index. Free-sprung balances have to be brought to time in the final stages by altering the moment of inertia of the balance by removing, adding or altering screws and/or by means of the quarter screws.

Frictional rest escapement An escapement that is always in contact with the balance.

Friction wheels See Anti-friction wheels.

Full-plate watch One in which all of the parts except the balance are located between two plates. An inconvenient arrangement that took a long time to disappear.

Fusee A grooved tapered pulley which when properly made evens out the torque output from the mainspring. When the mainspring is fully wound it pulls on a smaller diameter than when it is run down. The best watches had their fusees matched to their individual spring.

Fusee barrel The type of barrel that is used in conjunction with a fusee. It has no teeth, the great wheel being mounted on the fusee.

Fusee chain The chain that connects the barrel to the fusee. The making of fusee chains soon became a separate trade. It was invented in about 1635, and by 1680 was almost universally used in fusee watches.

Gadrooning Ornamentation found on the edges of late 18th- and early 19th-century watches. It is a sort of pie-crust ornamentation either made by hammering or casting.

Gate The name of the decorative piece covering the fusee stop-finger support.

Gathering pallet The pallet that gathers up the rack tooth by tooth whilst striking occurs.

Gimbals The freely swinging supports that are designed to keep a deck watch or chronometer horizontal regardless of the attitude of the outer case.

Going barrel A barrel that carries the great wheel on its periphery or on its arbor.

Going fusee A fusee with maintaining power.

Going train The train that has to do with the timekeeping side of the watch as opposed, for example, to the striking, repeating or musical train.

Gongs Strips of steel, usually circular, that go around the outside of the watch plates. Usually of round section but sometimes square. Invented by A. L. Breguet. They are struck by hammers.

Grande sonnerie Striking both the hours and the quarters at each quarter.

Great wheel The first wheel in the train and the slowest moving. Attached to the barrel or the fusee.

Greenwich Mean Time (G.M.T.) Mean Solar Time at The Old

Greenwich Observatory. Since Solar Time varies according to whether one is east or west of any reference point it is of little use in modern communities where travel and rapid communication are common. As a result, in 1880 G.M.T. was established by law as the official time for the whole of Great Britain. In 1884 Greenwich was taken as the prime or zero meridian from which all others are taken and time zones established around the world that relate to it. Only small countries can have just one time zone. The United States of America is so large that it must have five time zones each differing by an hour from that next to it.

Grey Parts in the 'grey' are not yet finally finished or polished.

Grisaille A technique in enamelling similar to 'scraper board' where a dark ground is laid down and covered with white enamel which is then scraped away to a greater or lesser extent to form a picture or pattern. This could be rendered more subtle by hatchings and by varying the tones of the white (either by using a spatula or a brush).

Guard pin The pin that is part of the safety action in the lever escapement. It is attached to the lever and is centrally disposed with respect to the notch.

Guilloché See Engine turning.

Hairspring The common name for the balance spring. A loose term that can refer to springs not associated with a balance, as in an electrical meter for instance.

Half plate A watch in which only the barrel, centre and third wheels are under the same top plate.

Half quarter repeater A repeating watch that sounds single blows for the hours, a double blow for each quarter hour past the hour and another single blow if another $7\frac{1}{2}$ minutes have passed since the last quarter.

Hallmark A mark applied by the Assay Office showing the place of assay, the metal purity, the year of assay and the sponsor's mark. Applicable in the U.K. to gold, silver and recently platinum.

Hand setting In early watches the hand was set by pushing it. Later the square on which the hand was mounted was extended so that the hand could be set by a key. Eventually the hand was set by means of the pendant button by pulling out the button or by moving out or pushing in a set piece.

Hanging barrel One which is supported at one end only, the top end; the other end of the barrel being in a clearance hole in the bottom plate.

Heart piece The heart-shaped cam which is operated by a lever to return a hand to zero.

Helical spring A spring which is in the form of a helix. The ends are normally incurved so as to obtain more uniform dilation of the spring as it works.

Hog's bristle A bristle used in the regulation of the balance or foliot before the introduction of the balance spring. The arm hit the bristle or bristles at some point before the end of its excursion, the exact position being adjustable.

Hole The bearing in which a pivot runs. May be of jewel, brass, bronze, lignum vitae, plastics, sintered material etc.

Hooke's Law The law that says that for a given material the relationship of stress to strain is a constant. Stress is force per unit area, and strain the extension per unit length. This is only true within the elastic limit of the material, that is within the limit where any deformation due to stress is not permanent after the load is removed.

Horizontal escapement Another name for the cylinder escapement.

Horizontal positions The positions in which a watch is horizontal, i.e. dial up and down.

Horns The extensions to the notch ends that form part of the safety action in a lever escapement.

Hour rack The rack that falls on to the hour snail and dictates the number of hours to be struck.

Hour wheel The wheel on which the hour hand is mounted, usually on a pipe extension. Normally the hour wheel rotates every 12 hours but sometimes it rotates in 6 or 24 hours.

Hunter A later case that has a solid cover to the dial. If the centre of the cover is pierced so as to show the middle half of the dial it is called a half-hunter.

Impulse The force applied to the balance and spring which makes good the losses due to friction and the fanning of the air.

Impulse angle The angle through which the balance moves during impulse is known as the impulse angle.

Impulse face The face through which the impulse is transmitted.

Impulse pallet The pallet through which impulse is transmitted.

Impulse pin The pin through which impulse is transmitted.

Impulse roller The roller through which impulse is transmitted or which carries an impulse pin.

Independent seconds A watch with two trains, one of which carries a seconds hand which can be stopped and started at will and whose rate of revolution is controlled by the other train. The independent train usually has its own minute and hour hands.

Index The piece that carries the index pins.

Index pins The pins that embrace the balance spring and can be moved along it for regulation purposes.

Intermediate wheel and pinion A wheel and pinion that lies between the great wheel and the centre wheel.

Isochronous The property of oscillating at a constant rate, despite the amplitude of oscillation. An isochronous balance will perform long arcs in the same length of time as short arcs.

Jacquemarts Figures that are moved by the watch mechanism usually on repeating watches.

Jewels The art of piercing jewels for use as bearings and applying them to watch work was perfected by Facio in 1701. At first the art was confined to England; it was not until later that it spread to the Continent.

Jumper A spring, or spring controlled piece, used to locate another piece.

Jumping Hours An hour hand that jumps forward once every hour. Breguet was fond of this arrangement in his repeating watches.

Jumping seconds A hand that moves forward in second jumps, despite the fact that the escapement moves forward at more frequent intervals, is said to be jumping seconds. Also a hand that each second completes one revolution of a subsidiary dial.

Karrusel A watch with a slowly revolving escapement. If it revolves in six minutes or less it is usually called a tourbillon.

Keyless watch A watch that can be wound and set without the need for a separate key.

Keyless winding Winding that can be affected without the need for a separate key. Early keyless winding watches still needed a key for hand setting. See Rocking bar keyless work.

Lepine calibre A watch movement in which the top plate is replaced by bridges and cocks. Introduced by J. A. Lepine in about 1770. This layout made possible a thinner watch.

Lever escapement The most successful of all the watch escapements, so called because a lever is interposed between the escape wheel and the balance. The balance first moves the lever to affect unlocking then the lever is pushed across by the escape wheel, inpulsing the balance in the process. When the pieces that the escape-wheel teeth act upon are jewelled the escapement is known as a 'jewelled lever'. If the pieces are pin-shaped, however, it is called a pin-pallet or pin-lever escapement regardless as to whether the pins are pieces of jewel or steel.

The escapement was invented by Thomas Mudge in about 1754 and subsequently improved over a long period until it became virtually standardised in its present form, the club-tooth lever.

Lever notch In a lever escapement the opening at the end of the lever which works with the impulse pin.

Lift The travel of the lever during impulse.

Ligne $\frac{1}{12}$ of the old French inch and equivalent to 2.25 mm.

Liverpool jewels Very large jewels fitted to Liverpool watches during the first half of the 19th century.

Locking The amount required for safety of action of an escapement. It represents the amount that the piece that is moved to unlock the escapement has to move before impulse begins.

Locking plate striking An earlier type of striking which has a locking plate or count wheel to determine the number of blows struck. Appropriately spaced notches around the rim of the wheel determine how long the striking train shall run before it is locked again. Unfortunately, once the hour has struck it cannot be struck again as with rack striking and as a result the striking can get out of phase with the hands. This will happen if for instance the striking train runs down before the going train and to rectify the situation may require specialised knowledge.

Lunette A rounded, slightly domed watch glass.

Mainspring The spring which provides the driving power for the various trains in watches, going, striking, repeating, musical etc.

Maintaining power A device for keeping a watch going when it is being wound where otherwise it would stop – as when it has a fusee. Usually Harrison's maintaining power but Arnold did fit sun and planet maintaining power in some of his early watches and chronometers.

Maltese Cross A wheel of that shape which forms part of stop work fitted to going barrels, this stop work being known as Geneva.

Massey lever escapement See Crank lever escapement.

Mean Time The average of all the solar days in the year is the mean solar day. This used as the basis of time is 'mean time'.

Middle temperature error The elasticity of the balance spring does not vary in the same way as the moment of inertia of the balance when the temperature changes. The error can be matched either at one or two temperatures which can be chosen at will. The choice is usually two temperatures, the extremes, which then results in errors everywhere between these two temperatures. This is called middle temperature error and can be overcome partially or completely by means of adding auxilliary compensation. By using a Guillaume balance middle temperature error can be avoided.

Minute repeater A repeater that repeats the hours, the quarters and then the minutes past the quarter.

Minute wheel and pinion See Motion work.

Modele deposé See Brevet.

Moon hand See Breguet Hands.

Motion work The gearing beneath the dial which causes the hour hand to revolve twelve times (or unusually twenty-four times) slower than the minute hand. It consists of the cannon pinion, the hour wheel, the minute wheel and pinion.

Movement The 'works' of a watch, that is a watch less case, hands and dial. For rough movement see Ebauche.

Musical watch A watch that has a separate mechanism that plays a tune on bells or a steel comb. The pinned barrel, as in a musical box, is sometimes used and sometimes a flat wheel with pins protruding from its face or faces.

Niello A process similar to Champlevé enamel but the depressions are filled with a black compound consisting of silver, lead and sulphur instead of enamel. May be on gold but is more usual on silver.

Nuremburg egg A misnomer due to bad translating of Uhrlein (little watch) as Euerlein (little egg). These watches were not egg-shaped but spherical or drum-shaped.

Offset seconds Small seconds where the seconds hand is not concentric with the hour and minutes hands and in addition not at the dial centre.

Oil sink A depression around a pivot hole to help retain oil and to make the amount that can safely be applied greater; surprisingly not introduced until about 1715 by Henry Sully. Early jewels had no sinks but these were soon included.

Oignon A popular name mainly applied to the large and bulbous French watches of the late 17th and early 18th century.

Open face An open-face watch that has no metallic cover to the dial. Although the time can be read without opening the front cover of a half-hunter it is still not termed open face.

Ormskirk A town in Lancashire, U.K. In the early 19th century a number of watches were made with a sort of Debaufre escapement and these are known as 'Ormskirk' watches.

Oscillation A constant and repetitious sequence of events. A pendulum and balance oscillate, as do a tuning fork and a quartz crystal, provided that some source of energy is available to maintain the oscillations, that is, to make good the various losses in the oscillating system.

Outer case The outer case of a pair-cased watch. Sometimes embellished, sometimes plain.

Overcoil The coil of the flat spiral balance spring that is raised above the body of the spring and is then shaped so as to follow certain rules. Invented by A. L. Breguet.

Pair case The standard form of the case for English watches between the mid-17th century and the end of the 18th century and beyond. The inner case contained the movement and was protected in its turn by an outer case. The outer case was often decorated in the style of the period; inner cases more frequently being plain. At the end of the period the outer case tended to be more plain.

Pallet Specifically the part through which the escape wheel impulses the balance. When this part is made of jewel it is called the pallet stone.

Parachute A shock-protecting device for the balance staff invented by Breguet. The balance endstones are not rigidly mounted but are held by long springy arms which can give under shock. The earliest form of shock protection.

Passing crescent See Crescent.

Passing spring Also known as the 'gold spring', because it is usually made of gold. It is a passing spring which allows the discharging pallet to pass the detent in one direction without moving it aside.

Pedometer watch This has different meanings. Sometimes taken to mean an early self-wound pocket watch whose self-winding weight oscillates at every step. Also a watch and pedometer combined.

Pedometer wind See Perpetual watch.

Pendant The part on top of the watch to which the bow is fixed. This can, by its form, help to date a watch. Latterly, the pendant carries the winding button in keyless watches, the winding stem passing through it.

Perpetual calendar A calendar mechanism that corrects the date according to the length of the months including the extra day in Leap Year. Usually shows day, date, month, these latter being on a four-year dial with Leap Year indicated.

Perpetual watch The English translation of the 'montre perpétuelle', the French term for a self-winding watch.

Pillars These locate, support and separate the plates or a plate, bridge, cock or potence. The earliest pillars were of square section, later, after about 1600 styles became diverse, Egyptian, tulip, pierced foliate, lyre shaped, square section baluster and then cylindrical.

Pillar plate The plate nearest the dial to which the pillars are fixed.

Pin barrel The barrel that carried the pins that play the comb in a musical train.

Pin pallet escapement See Lever escapement.

Pinion Usually the driven gear in watch work with six or more leaves, as the teeth are called.

Pinchbeck An alloy of zinc and copper, named after the inventor Christopher Pinchbeck in about 1730. It had a close resemblance to gold.

Pin work See Piqué.

Piqué or Pin Work. Pins of gold, silver or brass which are used practically to secure a covering to the outer case of a pair-cased watch. By arranging the heads in patterns they were also made to serve a decorative purpose. Leather, shagreen and tortoise-shell were some of the coverings in question.

Pirouette A staff with an integral pinion by means of which the balance could be given an exceptionally large arc.

Pitch The distance between teeth along the pitch line.

Pivot The bearing area of a rotating piece, usually, but not necessarily, at the end of the piece. Usually made of as small a diameter as is reasonable to reduce frictional losses.

Pivoted detent A chronometer detent which is supported by pivots, as against a spring detent.

Planting The process of determining the position of mating parts and making the holes in which they run or that take the jewels that form the bearings.

Plates The main members of the watch. The frame that locates and supports all the other parts. There may of course be only one plate.

Poise If the poise of a balance is perfect it will display no heavy point when supported on knife edges.

Pointed tooth-lever escapement See Lever escapement.

Poker hand The minute hand that usually matches the beetle hand, somewhat in the shape of a poker.

Positional error A difference in rate that results from a change in position is known as a positional error. It may be dial up to dial down or between dial up and the vertical position pendant up, and so on.

Potence (or Potance) A cock that lies between the plates.

Potence plate An old name for the top plate, because in verge watches it carried the potence that supported the lower pivot of the verge.

Pouzait lever escapement One of the first lever escapements with divided lift, invented by J. M. Pouzait in 1786. The escape-wheel teeth stand out from the wheel edge and these engage with a claw-shaped lever. The notch in the lever imparts impulse in the usual way to a steel impulse pin. The safety action was novel and used a ring with a gap on the staff that interacted with a pin on one arm of the lever. Thus, this pin would alternately be inside and outside the ring.

Pull wind See Pump wind.

Pump wind An early type of keyless winding. The watch is wound either by pulling or pushing a lever that passes through the pendant and terminates in a button. Massey and Viner both used this method.

Pulse piece A pin that projects through the case edge that is made to move against the bell hammer at will, so that repeating the watch can enable one to feel the hammer blows instead of hearing them on the bell.

Puritan watch Made between about 1625 and 1650, these watches were usually oval in shape and made of silver. As their name suggests, they were devoid of decoration.

Quarter rack See Hour rack.

Quarter repeater A watch that strikes the hours and the quarters at will.

Quarter screws The screws in a compensation balance that are adjustable so as to alter the moment of inertia and/or the poise. Because they are set 90° apart they are called quarter screws. The screws are either made a close fit in the threaded hole in the balance or else the balance is slit, so that the hole may be closed slightly so as to give a springy grip, as in the Glashütte balance.

Rack Toothed segment. See Hour rack.

Rack hook In striking or chiming work the lever that engages with the rack. It holds it and prevents it from slipping back, whilst not, however, preventing the gathering pallet from lifting the rack as striking occurs.

Rack lever escapement A form of lever escapement first invented by the Abbé de Hautefeuille in 1722. It was later patented in a more practical form by Peter Litherland in 1791 (patent 1830). The lever terminates in a toothed segment or rack which is permanently in mesh with a pinion on the balance staff. Made in large numbers in Liverpool in the early 19th century.

Rack striking See Hour rack.

Rack tail The tail of the rack which on contacting the snail stops the rack from falling further and thus dictates the number of blows to be struck.

Ratchet wheel A wheel which in conjunction with a click or ratchet has teeth so shaped that it can move in one direction only, or such that some change in the position of the click must occur for it to move in the other direction.

Rate The amount by which a timekeeper varies from the true time that it is intended to show. A timekeeper can have a large rate but providing it is uniform and predictable it can still be a perfect timekeeper – a fact not widely appreciated in the eighteenth century.

Recoil The backward movement of an escape wheel during the normal escapement operation. All truly successful watch escapements have recoil, however small, as this is utilised to provide 'draw'.

Recoil escapement An escapement in which there is recoil.

Regulation In pre-balance spring watches regulation was mainly done by altering the set-up of the mainspring. One way (to begin with) was by using a spring ratchet with a ratchet wheel on the arbor, but later also by utilising a worm and wheel drive. The worm wheel, which was mounted on the barrel arbor, usually had a graduated disc mounted on it so that one could see what adjustment had been made. If the helix angle of a worm is sufficiently small the arrangement is self-locking, so that it can be turned both ways with equal ease. A very early regulation arrangement was the use of the Hog's Bristle. Here the supplementary arc of the balance could be altered by varying the position of two upright hog's bristles against which the balance arms banked.

The ratchet set-up lasted until about 1640. The worm and wheel method (often called the tangent screw and wheel) only gave way when the balance spring was introduced. It gave way to Tompion's arrangement. The index plate now covered a gear that meshed with the segment of a larger gear that carried the curb pins. These pins can be made to traverse the spring by turning a square fixed to the index with the winding key. As the spring is effectively shortened or lengthened, so the rate changes. This was an elaborate arrangement and needed a special bridge to cover and locate the parts.

Regulator See Regulation.

Remontoire (or Remontoir) A constant-force device placed in the train near to the balance. It supplies the power to maintain the balance and is rewound at short intervals by the main source of power. Thus, the main variations due to the gear train and the running down of the mainspring are eliminated.

Remontoire escapement A remontoire device that is mounted on the escape wheel or next to the balance. Usually called a constant-force escapement.

Repeat A watch repeats if one can make it strike at will.

Repeater There are many types of repeater. Quarter, half quarter, minute, five minute and half ten minute. The watch is made to repeat by pressing on a plunger, pushing a push piece or pulling round a slide. The hammers may strike a bell, on gongs, or on a block, this latter being known as a dumb repeater. Repeating work was invented by Daniel Quare in the 1680s.

Repoussé A pattern raised by hammering from the back so that the scene is in relief.

Resilient escapement A form of lever escapement in which when banking occurs the blow on the impulse pin is taken by a non-rigid part of the lever, or in which the escapement is so arranged that recoiling the escape wheel serves the same purpose.

Robin escapement A lever chronometer escapement made by Robin in the 18th century. Locking takes place on the lever and the impulse is given direct by the escape wheel to the balance. It is single beat.

Rocking bar keyless work The action of the keyless winding mechanism usually found in later going barrel English watches. A bar is rocked by means of a push piece to move gears into mesh for handsetting and out of mesh for winding. On releasing the push piece the bar moves back under the action of a spring into the winding position.

Roller The part of the balance staff that receives impulse, or carries the pin, jewel etc. that receives impulse.

Roskopf G. T. Roskopf manufactured the first inexpensive watches in 1867. They had a pin-pallet escapement and were keyless pocket watches. The name Roskopf came to mean a certain type of watch and the name is often inaccurately applied.

Ruby-cylinder escapement Because the steel cylinder of a cylinder escapement was found to wear, some cylinders in high quality watches were made of ruby. Breguet especially used many ruby cylinders and devised a special sort that hung down below the bottom pivot of the balance.

Run-to banking In a lever escapement the extra movement of the lever needed above the bare movement required to allow the escapement action to occur. The extra movement is required for safe action.

'S' balance An early form of compensation balance used by John Arnold. 'S'-shaped bimetallic strips caused weights to move nearer to or further away from the centre of the balance as temperature changes occurred.

Safety action See Lever escapement.

Safety roller See Lever escapement.

Safety dart See Dart.

Savage two pin A form of lever escapement developed by George Savage. Introduced about 1814.

Savonnette A watch with a front cover to protect the glass.

Scape wheel The wheel that delivers impulse directly to the balance or through an intermediary. It is the last of the train wheels but does not mesh with another pinion. Properly escape wheel.

Secondary compensation Compensation that is additional to the main compensation and that corrects small residual errors.

Seconds train A train so designed that one of its wheels rotates once a minute, and can therefore have a seconds hand mounted on it.

Seconds wheel and pinion The wheel and pinion in the train that rotates once a minute.

Secret signature Used by Breguet to protect himself against forgeries. Has to be looked for carefully and is usually near the XII. Other makers subsequently adopted the idea.

Secret spring The fly and lock springs of a hunter case.

Segment weights The weights that are slid along the arms of a compensation balance to adjust the amount of compensation. Called segment weights because of their shape.

Self-compensating spring Somewhat of a misnomer but nevertheless a convenient term for a spring that has a small thermal coefficient of Youngs modulus that can be manipulated so as to compensate for its own dimensional changes and those of the balance. For complete compensation there must still be some adjustable device attached to the balance, or else the balance must be of the Guillaume, or of the ovalising type.

Self-winding A watch that is able to wind itself through the movements of the wearer. See Perpetual watch.

Set up Since the bottom diameter of a fusee cannot be infinitely large it cannot give the correct torque when the mainspring is completely run down. In practice, the bottom diameter is dictated by the size of the great wheel and the need for a maintaining ratchet (or winding ratchet in early watches). Therefore, the mainspring must be set up so as to give correct torque at this diameter and an equal torque when the mainspring is fully wound achieved by a reasonable size of the smallest diameter at the top of the fusee. Working out the shape of a fusee is a complex matter and is seldom done other than empirically or by virtue of experience.

S.G.D.G. See Brevet.

Sidereal Time Time as given by the stars. This is uniform in a way that solar time can never be. Lack of uniformity in Sidereal

Time is so slight that only when quartz clocks came into use was it established.

Single-beat escapement An escapement such as the chronometer, where impulse is given when the balance is travelling in one direction but not the other. Usually considered to be a drawback in an escapement as increasing its liability to set.

Single-plane escapement An escapement that works in one plane as with a lever escapement.

Single roller See Lever escapement.

Six-hour dial An attempt to read to the minute with a single hand was made at the end of the 17th century by having a six-hour dial. This can be divided at two-minute intervals. Of course, to tell the time one needed to know which six-hour period one was in.

Skeleton dial A dial that is cut away so that one can see the work beneath it.

Skeletonised movement A movement cut away so as to disclose as much of the work as possible. Taken to incredible extremes in some modern work.

Snail A cam-shaped piece usually dictating the number of blows to be struck.

Solid banking Banking that is provided by part of the bottom plate as opposed to a pin. See Banking.

Solar Time Time as indicated by the sun, e.g. as on a sun dial. Of little practical use in modern life, but early clocks were checked against sun dials by using the equation of time and local time which was used until the introduction of Greenwich Mean Time.

Souscription Breguet's inexpensive watch ordered in advance by the customer. Characterised by a barrel at the centre and a single hand through which the winding square passed.

Spotting A finishing method consisting of a series of overlapping spots arranged in concentric circles usually in watch work. Applied to a previously polished surface.

Split seconds A chronograph watch with two centre seconds hands that can be used to time two distinct events. One hand may be stopped, then if wished the other. When next the plunger is pressed the second hand returns to its position over the first, or both may be returned to zero, or both may continue together.

Spring barrel A barrel containing a mainspring. It is not clear when it was first invented but probably before 1450.

Spring detent The detent of a chronometer escapement that is supported and located by the spring that returns it to the locked position. Preferred by the English to the pivoted detent although possibly not with justification.

Stackfreed The device fitted to early watches to help to equalise the force of the mainspring regardless of its state of wind. A cam, mounted on the barrel stop work, is so shaped that a strong spring that follows it robs the mainspring of power when the watch is fully wound and assists the mainspring at a later stage. A crude device that was soon supplanted by the fusee.

Standing barrel A barrel, supported at its lower end only.

Star wheel A wheel with triangular-shaped teeth that can be indexed and is located by a jumper. Often carries a striking snail.

Stones Another term for jewels.

Stop work The arrangement fitted to a fusee or to a going barrel to limit the number of turns of winding. Always at the top end to prevent the spring being fully wound, and with going barrels usually limiting the number of turns by also stopping the barrel before it is unwound.

Straight-line lever A lever escapement in which the pivots of the balance, lever and escape wheel lie in a straight line. Almost invariably (there are exceptions) the layout of modern lever watches.

Stud The outer termination (or top termination with a helical spring) is either pinned into, clamped into or latterly sometimes cemented into a stud. This in turn is then screwed or clamped to the balance bridge or cock or the top plate.

Sugar tongs An early compensated curb-pin device shaped in fact like latter-day washing tongs. Both blades of the tongs are bimetallic and carry a curb pin at their termination.

Sun and moon dial A popular type of dial at the turn of the 18th century. A twenty-four-hour dial carries both a sun and moon. One indicates the hours of darkness from 6 p.m. to 6 a.m. and the other, the hours of light, 6 a.m. to 6 p.m. Doubtless more useful in the tropics than in temperate latitudes!

Sundial Time See Solar Time.

Supplementary arc The arc that a balance describes over and above that minimum required for escapement action. Probably the greatest supplementary arc is found in pirouettes, since the action need not be confined as with more conventional escapements, the action in fact being only limited by the power that can be put in, and by the necessity to prevent the hairspring coils from touching.

Surprise piece A device fitted to the snails of repeaters to prevent incorrect striking just after the hour or after each quarter.

Sweep seconds See Centre seconds.

Swing wheel An old term for the escape wheel.

Swiss lever escapement The term loosely used to indicate the club-tooth lever escapement, to distinguish it from the English or ratchet-tooth lever escapement.

Table roller The roller of the lever escapement on which is mounted the impulse pin.

Tact A montre à tact has a stout hand fitted over the back or the dial of the watch, which can be moved in a clockwise direction, until it is stopped by the mechanism inside. The time is then read off by feeling the position of the hand relative to touch pieces on the case edge. These pins are in the hour positions. Breguet made a number of these watches. They are a less expensive alternative to a repeater.

Taille douce Fine line engraving.

Tambour case An early type of drum-shaped case with a hinged lid.

Tangent screw A worm used in conjunction with a worm wheel for setting up the barrel (see Regulation). After 1675 this worm was moved inside the plates, before this it was mounted on the top plate.

Tavan lever An early type of lever escapement notable for being the first to have divided lift.

Temperature compensation Since metal (or most metals) change dimensions and stiffness with changes in temperature, some sort of compensation has to be affected if the balance and spring are not to change timekeeping characteristics. From the first attempts, two centuries were to pass before the problem was solved in a completely satisfactory manner.

Terminal curves Called incurves when discovered by Arnold, these were first applied to helical springs and helped to make the spring expand and contract uniformly. Subsequently applied by Breguet to flat spiral springs the mathematics of the theoretical curve was finally worked out by Phillips. The object, in theory, of the terminal curve is to keep the centre of gravity of the spring on the balance axis as the spring dilates and contracts. In fact, for close rating these curves are only the starting point and normally need to be modified to achieve close positional rates. Also the point of attachment of the inner end of the spring has to be controlled for best results.

Third wheel and pinion Normally the wheel and pinion next to the centre wheel and pinion which in turn drives the fourth wheel and pinion.

Three-quarter plate The top plate of a three-quarter plate watch covers the barrel, centre wheel, third and fourth wheels, but not the escapement.

Timepiece A timekeeper, pure and simple, that does not strike.

Timing in positions See Adjusted.

Timing screws See Quarter screws.

Timing nuts See Quarter screws.

Tinted gold Also called coloured gold. Gold can be produced in various colours according to the metals with which it is alloyed. This fact is used to provide golds that can be decorative because of colour and the way in which this is used. Often called four-colour gold and used for dials and sometimes cases, although only three or even two colours were used on many occasions.

Tipsy key A key designed by Breguet that can wind in one direction only; it free wheels in the other direction and then no damage can be done by winding in the wrong direction.

Tompion regulator See Regulation.

Top plate The plate that is on top if one considers the plate near the dial as the bottom plate.

Torque A turning moment. It is the product of force times distance, and its units in horology are usually grams/centimetres.

Touch pins Pins were usually inserted in the dials of 16th-century watches at the hour positions so that the time could be established in the dark. The single hands were very sturdy and not likely to be damaged by feeling for them.

Tourbillon A. L. Breguet patented his 'tourbillon regulator' in 1801. This was a carriage that revolved at one-, four- or six-minute intervals and that carried the escapement complete. This nullified the vertical positional errors.

Train The wheels and pinions that are used to connect the barrel to all the other moving parts in a watch. It may be the going train, the repeating train, or the musical train etc., according to its purpose.

Trial number A number arrived at by multiplying factors determined during trials in such a way that this number indicates the excellence of the timekeeping performance.

Triple case A watch, usually made for the Turkish market, that had three cases instead of what was at that time, in England at least, the more normal two – the pair case.

Tripping If the amplitude of the duplex or chronometer balance becomes too high the escapement can be unlocked a second time and a second tooth can escape during the one excursion of the balance. This is called tripping.

Turkish market watches Large quantities of watches were made for the Turkish market around the end of the 18th century and the beginning of the 19th.

Two-pin escapement See Savage two pin.

Two-plane escapement An escapement that is in two planes such as the verge.

Under dial work The work that lies between the dial plate and the dial. Also called 'cadrature'.

Up-and-down dial A dial showing the state of wind of the mainspring. Usually only found in higher class work.

Verge See Verge escapement.

Verge escapement Sometimes called the crown-wheel escapement because the escape wheel somewhat resembles a crown. The teeth are roughly triangular in shape and protrude from the face of the wheel. There are usually an odd number of teeth in the wheel (eleven or thirteen usually in watches) and one tooth is always pushing on one of the two 'flags' on the balance staff – called the verge.

Since the verge crosses the wheel, and one flag is on the opposite side of the crown wheel to the other as the second flag is picked up by the crown wheel, the other flag is moving in the opposite direction and the crown wheel is recoiled.

Vertical positions The positions when the balance is hanging vertically as opposed to being horizontal. Normally the four quarter positions are designated pendant up, pendant right, pendant down and pendant left. The horizontal positions are dial up and dial down.

Virgule escapement An escapement of the horizontal type that followed the cylinder escapement introduced in about 1750. The balance staff carries a pallet shaped like a comma. Upstanding pin-shaped teeth on the escape wheel slide up the part of the pallet that corresponds to the inside curve of the tail of the comma to give impulse. When drop occurs the next tooth falls on to the top of the comma, subsequently dropping into the notch formed where the dot meets the tail. When the balance reverses, the pin eventually unlocks and runs up the tail again giving impulse. The escapement is thus single beat.

Volute balance spring The flat spiral spring.

Wandering hour dial A dial without hands. The hour shows through a hole in a semi-circular slot and moves around this slot, its position indicating the minutes against a semi-circular minute dial that lies outside it.

Warning The operation of the striking train that takes place a few minutes before striking actually occurs. The rack is allowed to fall and settle, the train runs slightly until it is held on a pin on the warning wheel, so that all is in readiness for the exact moment of release of the warning wheel when the time has come to strike.

Watch papers These were originally used to protect the inner pair case from the inside of the back of the outer, which is where they were placed. Towards the end of the 18th century they also served as advertisements and often carried sentimental rhymes or sometimes the equation of time. Watch papers (a loose term; they were made of a variety of materials) are collected in their own right – a shame, since they constitute part of the history of the watch and would be better left where they were.

Waterbury long-wind watch This was a series of watches characterised by the elegance of the design and by the small number of parts – this latter for cheapness. They had a duplex escapement and also an exceptionally long mainspring and took a long time to wind.

Winding square The square on the barrel arbor or fusee by means of which the watch is wound. This square may be internal when the designer wishes to reduce the overall thickness of the watch.

Winding stem The shaft which carries the winding button. See Keyless winding.

Wolf-tooth gearing Gearing which drives in one direction only may have a special tooth form called wolf tooth because of a fancied resemblance in shape. Although commonly met with in keyless wheels, it has been used throughout the gear train in some rare watches.

Worm wheel See Tangent screw.

Z balance An early form of compensation balance designed by John Arnold.

Description of the Colour Plates

Plate 1 – This 16th century gilt metal alarum watch is probably German. It is pre-balance spring with balance and has a verge escapement. The lid which originally covered the watch has been cut away and undercut to take a glass – glasses were unknown at this period. The alarm setting disc can be seen at the dial centre. The hand is not original. The movement is steel with rectangular pillars and open mainsprings. The diameter is approximately 64 mm. The watch originally had a stackfreed, now removed.

Plate 2 – This Dutch watch is by Daniel Van Pilcom of Amsterdam. The case is silver and gilt made in the form of a fritillary flower. The fusee (with gut) movement is pre-hairspring and has a silver pinned cock. The pillars are Egyptian. There is ratchet set-up. The single hand registers on a gold chapter ring on an engraved dial. The case is a mere 30 mm long (without pendant). The view is of the case back. (c. 1620)

Plate 3 – Silver Skull Watch. Made by J. C. Vvolf, circa 1620. These rather macabre watches were made to remind people of the fleeting nature of life. The dial of the watch is revealed by opening the lower jaw. Verge escapement, pre-balance spring.

Plate 4 – This Swiss astronomical watch was made circa 1620 by Gaspard Girod à Geneve. It is verge, pre-balance spring with a balance. It is a fine example of its type. Note the screwed cock. The barrel set-up is with ratchet and ratchet wheel both of blued steel. The calendar indicates the following: Age and phase of moon, month and sign of Zodiac. The four quarters of the day are also indicated. The case back was originally crystal but this has been replaced with silver.

Plate 5 – This French verge watch is in an oval gold case pierced, enamelled and set with jewels, the back being emerald glass. The dial face is engraved on white ground, with decorations in translucent enamels. At the dial centre there is a landscape in coloured enamels. (c. 1620)

Plate 6 – This Swiss watch has a verge escapement. The case is crystal with a decorated enamel rim and bezel. The scene on the dial shows a bridge over a river with a fisherman in the foreground. The movement is inscribed J. Sermand à Geneve. (1595–1651)

Plate 7 – This crucifix watch is French. The case is embossed silver and gilt. The movement is verge pre-balance spring with balance and bears the inscription O. Tinelly. (1630–35)

Plate 8 – The case of this watch is circular and is of gold, painted with enamel flowers in relief – roses, tulips and pansies etc. on a black ground. The lid has a border and a centre of diamond rosettes in gold settings. Inside the lid there is a figure in grisaille. The dial is enamel being a landscape in colour with two figures. On the top plate is D. Bouquet Londini but the case probably came from Geneva. (1630–40)

Plate 9 – This French verge watch is very small for a watch of this period. (However, there is a smaller round watch in the Museum at La Chaux de Fonds in Switzerland of only 12 mm in diameter.) It is pre-balance spring with ratchet and ratchet wheel set-up for regulation. The case is gold and also serves as the dial. The chapter ring is white enamel and the case blue. The movement is inscribed A Bretonneau à Paris. (1638–43)

Plate 10 – This German watch is a rare item and has a pendulum balance with a straight hairspring. Made by George Seydell in the mid-17th century. The straight hairspring is now unfortunately missing. The case has a silver gilt band, silver bezels and is glazed front and back. It is an astronomical watch showing day, date and month, and with a lunar dial. The dial has silver chapters on an engraved gilt ground. This watch illustrates the interest engendered by the introduction of the pendulum into clocks.

Plates 11a, 11b and 11c – This English watch is verge, the movement being inscribed Isaac Pluvier Londini (1614–65). The case is gold and is enamelled. The dial has Roman numerals and a floral centre. There is a pastoral scene inside the front cover. This watch enamelling is exceptional because of the delicacy of the work. (c. 1650)

Plate 12 – This watch is painted enamel and on the back shows Venus and Adonis and is circa 1648. The movement is signed Estienne Ester.

Plate 13 – A French verge watch, dated 1630–40, made by B. Foucher. Scenes of the Amazons are shown in the enamelling which is attributed to Jean or Henri Toutin.

Plate 14 – A beautiful example of miniatures in enamel taken from the edge of the case of a French verge watch. The case is signed by the enameller André Père et Fils. The scenes are of houses in lakeland settings. Other views of this case are shown in Figs. 6 and 12. Note the dial with cartouches. The movement is inscribed Pierre Martin c. 1695.

Plate 15 – This Swiss clock-watch with alarum was made by Jean Baptiste Duboule of Geneva (1615–94). He was a master engraver as well as being a watchmaker so that he would have been responsible for the lovely dial which has phase and age of moon, sign of the Zodiac, date, rising and setting of the sun etc.

Plate 16 – This German clock-watch with alarm was made for the Turkish market. It has a single case of gilt finish, single hand, silver and gilt dial with Arabic numerals. Note the worm set-up for regulation purposes. The watch is pre-hairspring. (c. 1680)

Plate 17 – This French watch was made by L. Vautyer in the 17th century. It is a verge watch. The body of the case is enamelled with raised and pierced gold and enamel decoration.

Plate 18 – This German watch has a rock crystal case in the form of a snail. It is a pre-balance spring verge watch with fusee and chain. Made by Wilhelm Peffenhauser of Augsburg. The dial is silver with Roman numerals and there is a single hand. The dial has an engraved brass surround. (c. 1650)

Plate 19 – A pair-cased gilt metal watch by Tompion London – the 'father' of English watchmaking. It is a verge watch with alarm, 'onion' style. The concentric alarm dial has enamel cartouches. It has a single hand with bold roman numerals on the outer enamel chapter ring and dates from the last third of the 17th century.

Plates 20 and 21 – A French verge watch with pendulum cock. It is a clock-watch. The count wheel can be seen top right in the movement shot. The index-wheel regulator is top left. The bell can be seen in the case bottom. The silver case is pierced and engraved. A silver chapter ring is mounted on a gilt and engraved dial. The movement is inscribed Fardoil à Paris. (c. 1680)

Plate 22 – This watch although inscribed Barraud London No. 1419 is of Swiss manufacture. It is a verge watch with centre seconds. The enamel dial has a rustic scene and the sails of the windmill revolve, being mounted on the escape pivot. The pair cases are metal, gilt with paste-set bezels and enamelled back. (1756–94)

Plate 23 – This English watch, made in the late 17th century, has a verge escapement and silver wandering hours dial with Windmills London engraved in a cartouche. The movement is inscribed Jos Windmills 0717. It is pair cased in silver, the outer being engraved with the arms of King William III.

Plate 24 – This verge watch is by Fr. Stammer London and is numbered 699. It has a double six-hour dial. This was developed to try to give a single hand the same accuracy of reading as two hands. The outer ring is divided into two-minute intervals so that the time can be read to a minute. One of course had to know the time within six hours, or confusion could arise, say, for instance, between 4 and 10 o'clock. Silver Champlevé dial pair cases the inner silver. The outer is silver metal covered in shell and inlaid silver. (c. 1700)

Plates 25a, 25b and 25c – This English watch is double dialled, one dial has I–XII twice and shows the months with coloured signs of the Zodiac. The other side shows hours, minutes and the date. The dial is inscribed David Pons. The gold case is glazed both sides. There are two worm and wheel drives in the movement which has a cylinder escapement. (c. 1770)

Plate 26 – This English watch is so large that one wonders if one can call it so. It is cased as a chronometer but in all its details it is nothing so much as an overgrown watch. Made by George Margetts (1748–1908) it gives the following information: Mean Time hands at centre. The state of tide at eight ports also on this centre dial and indicated by the surrounding twenty-four hour dial. This latter dial also carries the moon hand. The moon indicates its position in the Zodiac and its declination on the outer circles on the large dial. These show from the inner to the outermost, declination degrees, sign and degree of the Zodiac, and date and month of the solar year. The sun hand indicates the date and as with the moon, its own declination and position in the Zodiac. The circular frame separating the outer circles from the constellations has the moon's latitude above or below the indicated declination engraved upon it. The dragon's-head pointer also indicates the position of the moon's eclipse nodes in the Zodiac. The outer curved frame represents the observer's horizon, and the space between it and the inner curved frame, the twilight period. All parts rotate clockwise except for the horizon frame and the Mean Time dial, which are stationary. Also through the hole in the twenty-four-hour dial is indicated eclipses of the sun and moon. Between eclipses the age of the moon engraved on the sun hand disc (1–29½) can be observed each day through this same hole. All of this is accomplished with just sixteen gears in the motion work. True watches on exactly the same plan were also made by Margetts. (c. 1780)

Plate 27 – This is a view of the back of the case of a Dutch watch with verge escapement. It is a quarter repeater with gold champlevé dial of typical Dutch design. The inner case is plain gold, the outer repoussé. Both cases are pierced. The movement inscribed B Van Der Cloesen, Hague (1688–1719). (c. 1710)

Plate 28 – This is an English watch case of gold, enamelled blue and white, with a portrait of Arabella Hamlyn, 1774. It was enamelled by Henry Spicer (1743–1804) who was appointed enameller to George, Prince of Wales.

Plate 29 – This English watch by John Arnold is full plate with spring detent escapement and is half quarter dumb repeater. The movement is inscribed John Arnold London Invt et Fecit No. 21/68. The movement was originally pivoted detent but has been converted (by Arnold probably) to spring detent. It has a double S balance with helical spring. The movement is jewelled throughout including the fusee arbor. The case is gold hallmarked London 1780, the dial enamel, hands gold.

Plate 30 – This French watch is one of Breguet's 'souscription' watches which were Breguet's inexpensive line. Ordered before they were made on a subscription basis they were as simple as they could be. The barrel is at the centre of the watch, the winding square can be seen in the centre of the single hand. Note the quiet elegance of the Arabic numerals and the fine hand. The number of the watch is 2267. It was originally numbered 267 and this number appears against the secret signature just below the 12 o'clock. This is not visible in the photograph. This watch has a ruby cylinder escapement with a three-armed plain balance. The regulator has a compensation curb. The movement is three-quarter plate. The case is plain silver with gold bezels and band 62 mm diameter. 1798.

Plate 31 – This watch shows the early enamel dial made up from cartouches. The case is a piqué case.

Plate 32 – This watch is unusual, inasmuch as a portrait is painted inside the front cover.

Plate 33 – This watch shows really superb enamelling where the flowers almost seem to stand away from the surface.

Plate 34 – This Swiss anonymous musical watch is of the Fleurier type, and strikes on five bells. The case is gold and has what are apparently forged English assay marks. The dial is enamel. The cuvette is very beautiful being pierced and engraved with an urn of flowers. The bezel, pendant and bow are decorated with split pearls. The back is of translucent blue enamel on a guilloché ground and is also decorated with split pearls around the edge, with a reticulate motif of split pearls at the centre. (Early 19th century)

Plate 35 – This anonymous Swiss watch is a quarter repeater with automata. Two cupids ring bells at the upper part of the dial and two figures operate a grinding machine in the lower part. The escapement is cylinder. The case is gold. (Early 19th century)

Plates 36a and 36b – A Swiss watch with verge escapement. The gold case is decorated all over with grains, filigree and amethysts. The dial is gold with numerals on oval plaques. The movement is inscribed Fx Pernetti à Geneve No. 13601. (c. 1810)

Plate 37 – This watch, originally made by John Arnold in 1774 or 1775, was fitted with a tourbillon mechanism by A. L. Breguet, and presented to John Roger Arnold in 1809 as a token of esteem from one famous maker to another. This esteem was mutual since at one stage there was an exchange of sons so that each could serve some time in the other's establishment. The silver plate attached to the top plate can be translated as follows: The first tourbillon regulator by Breguet incorporated in one of the first works of Arnold. Breguet's homage to the revered memory of Arnold. The escapement is spring detent of the Peto cross detent type, with a steel escape wheel. Breguet's records show that over the course of a year fourteen different workmen attended to the modification of the movement, to the making of the tourbillon, and the casing and dialling of this watch. This watch has sun and planet maintaining power similar to that in Arnold's No. 1 marine chronometer and in other pieces of his manufacture. As far as is known Arnold was the only one to use this type of maintaining power in watches or chronometers, which he only did in early examples. The case is silver, engine turned and is in a presentation case not intended for wear.

Plate 38 – This Breguet watch, No. 971, shows one of his attempts at escapement improvement. The watch is a repeater and has a vertical wheel natural lift escapement. This is a variety of the resting wheel or dead beat verge. However, the function of locking is transferred to another wheel mounted co-axially with the impulse wheel which is locked by a lever with upright pallets that has the normal fork and double roller arrangement. This watch survives as a movement only. (c. 1810)

Plate 39 – This Swiss watch is a quarter repeater with skeletonised movement and has automata. The dial has coloured gold figures and foliage. Made by Meuron and Co. the movement has a cylinder escapement. The case is gold. (1800–25)

Plate 40 – This club-tooth lever watch carries a well-known English name, Robert Roskell, Liverpool, but actually was imported from Switzerland and is of the Fleurier type. The enamel dial is later and has both a thermometer and a compass put into it. (c. 1836)

Plate 41 – This is a French cylinder watch with repeating. Two gold figures strike bells which are mounted on the dial. This is unusual, as the figures are normally only pictorial and do not strike real bells. The offset dial shows the hours and minutes, the whole is mounted on a blue enamel ground. The movement is fusee, and is inscribed Selliard Aine à Paris No. 4050. (1812–25)

Plate 42 – These watches show different styles seen in cylinder watches. The lower one is elaborately enamelled and engraved and has a shaped case, while the top watch has been designed with simplicity in mind.

Plate 43 – This watch is a musical (two tunes) quarter striking clock-watch. It has an engine turned silver dial with the age and phase of the moon. The going train is powered by a going barrel and has a cylinder escapement. The name Courvoisier appears on the dial. The gold case is also engine turned. (c. 1815)

Plate 44 – This French watch is an independent seconds watch. The dial is silver, the case gold and both are engine turned. The escapement is ruby cylinder, with steel escape wheel. (c. 1815)

Plates 45a and 45b – This watch made by Ls George & Ce. Hr du Roy à Berlin, is a quarter repeater, and shows the day of the week and the date. It has a Pouzait-type lever escapement and is centre seconds. The large balance covers nearly all of the top plate! The case is silver, the dial enamel. (c. 1815)

Plate 46 – This Breguet watch has a cylinder escapement and is 'à tact'. It has a gold dial and gold case. The movement is inscribed Breguet Horloge de la Marine Royale No. 3877. (c. 1820)

Plate 47 – An English clock-watch bearing the name French, Royal Exchange No. 4458. It has a duplex escapement. It is also a quarter repeater. The striking train has a most unusual refinement. A cam on the striking stop work alters the fly-pinion depth as the mainspring runs down, in such a way as to regularise the speed of striking. The balance cock and barrel bridge are beautifully engraved. The dial is silver. (c. 1820)

Plate 48 – This watch is probably French. It has a cylinder escapement and is very flat whilst still being hunter cased. It has a gold niello case and gold niello dial. Note the matching chain and key. Moving discs show the hours and minutes. (First half of the 19th century)

Plate 49 – This English watch was made by John Moncas of Liverpool and displays magnificently the 'Liverpool jewelling', with large, relatively clear jewels. The balance is gold and the escapement is of the later Massey type. The engraving is bold and striking, and the gilding rich. (c. 1820)

Plate 50 – This Swiss watch has a Tavan lever escapement and compensation curb. It is also a repeater. The dial is enamel and also shows the date. The movement is inscribed Jn. Ge. Remond a Geneve. (c. 1830)

Plate 51 – This English watch has a ratchet-tooth lever escapement. The movement is inscribed Ulrich & Co. Cornhill London No. 62. It is a half plate with exceptionally high numbered pinions and wheels. They are as follows: Fusee 104, Centre Pinion 14; Centre Wheel 112, Third Pinion 14; Third Wheel 105, Fourth Pinion 14; Fourth Wheel 96, Escape Pinion 12; Scape Wheel 15. The escapement has diamond endstones. The balance is bimetallic with three arms. (1828–32)

Plate 52 – This watch is inscribed Hunt and Roskell 156 New Bond St. London No. 10514 and is absolutely ravishing. However, it was probably J. F. Houriet (1743–1830) who made it. The watch is a one-minute tourbillon with a 21,600 beat rate – unusual in a tourbillon, especially at that date. The polishing on the carriage is as superb as anything ever seen. The escape wheel is gold. The movement is keywound fusee with continental stop work. The fine mainspring is signed Bandelien Frere. A brake for the train operates on the third wheel. The balance is unusual. It is bimetallic with special screws in the segment weights for fine adjustments of the compensation. It is free sprung with a spherical spring. The escapement is Earnshaw-type spring detent. An additional complication is the inclusion of a thermometer. The case is gold, the dial enamel. (1836)

Plate 53 – This is a ball pendant watch with enamelled case. The watch is wound by revolving one half of the ball one way relative to the other. Reversing the direction sets the hands. These watches sometimes have cylinder and sometimes lever escapements.

Plate 54 – This Viennese watch is a copy of a 1650 style watch. The case is crystal front and back, set in enamel bands. The dial is also enamelled and there is a single hand. Whether these were intended to be fakes is by no means certain. (c. 1840)

Plate 55 – This English watch has a right-angled lever escapement (club tooth) with split seconds. Made by E. J. Dent, it is marked: compound movement chronometer makers to the Queen, London 15430 Patent. It is an early example of a split seconds watch and works on an unusual principle. It has a bimetallic balance with gold timing and quarter screws, with an overcoil spring pinned to a

Breguet-type stud clamped under a steel plate. The movement is of Lepine construction. (c. 1846)

Plate 56 – This has been described as a Japanese paper-weight clock. This little gem has also been described in Japanese papers as a 'doctor's clock', and although it might well have been it really comes under the heading of a watch. The confusion really arises because it is modelled on the well-known Japanese Pillar Clock. However, this watch is spring driven thus making it portable, and it also has a balance and spring. The outer case is of Shitan wood (lid has been removed for photography and the watch removed from its outer case). As the watch is wound, the hour indicator, which is fixed to the fusee line, moves to the top of the hour scale. As the watch runs, the pointer moves towards the bottom of the scale indicating the hours. The hour marks are adjustable. (Probably 19th century)

Plate 57 – This Japanese Inro Watch is mid-19th-century with verge escapement. The inro, or medicine box, is eminently suitable for housing a portable timepiece and this example shows one way in which this was done. The kimono provides no receptacle for carrying a pocket watch, but of course the inro was attached to the girdle by its netsuke and ojine. The watch is really a miniature clock with rectangular plates and striking mechanism. The bell is visible through the slot near the top of the box. The box is of Shitan wood adorned with maple leaves, chrysanthemum flowers and butterflies. The ojine is decorated with a spray of leaves and coral fruit and has a mother-of-pearl cartouche, with the signature Furuya. The striking is based on the European system. The dial is gilt with moveable silver and steel numerals.

Plate 58 – This Japanese watch has a verge escapement. There is a twenty-four-hour dial with adjustable numerals striking on Japanese notation. Dial engraved, gilded metal and steel chapters which are moveable. Fixed hand. Silver open-face case back containing bell pierced and engraved. Whole in rectangular wooden case with sliding back. Diameter 81 mm. (Probably 19th century)

Plate 59 – This Swiss cylinder watch has a movement made completely of steel! The style of the movement is Bovet. The watch is centre seconds, the dial enamel, the case silver. (c. 1860)

Plates 60 and 61 – In 1825 a new era began for the Swiss watch trade in China. This was started by Edouard Bovet who had gone to China to represent a London firm. However, he soon became independent, and organised a trade that became of such proportions that the Bovets of Fleurier Switzerland became known as the Bovets of China. The calibre was distinctive, as can be seen in the illustration, and became known as the 'bovet'. It was characterised by the layout of the bridges and the beautiful engraving. After 1842 a few Neuchâtel firms Vaucher, Dimier Vrard, and later Juvet and the firm of Voumard and Courvoisier (both the latter of La Chaux de Fonds) undertook production of similar watches. (c. 1860)

Plate 62 – A Swiss watch, Systeme Roskopf. This was the first attempt to make a watch for the working man, a task undertaken by Georges Frederick Roskopf (1813–89). The watch has a pin-lever escapement and is keyless (rocking bar system). The case is nickel, the dial enamel. (c. 1870)

Plate 63 – This Swiss watch has a pivoted detent chronometer escapement. The movement is inscribed Albert H. Potter & Co. Geneva. Pat Oct 11 '75. Potter was a superb craftsman who left America to settle in Geneva in 1876, and was actually credited by the Swiss with raising the standard in the area! (c. 1876)

Plate 64 – An early American Waltham full-plate lever watch. This watch is only jewelled at the balance and pallets. Note the works safety pinion. The centre pinion is screwed on to the centre arbor and would unscrew if the force became too great upon a mainspring breaking in the going barrel. (c. 1880)

Plate 65 – This Swiss one-minute tourbillon with lever escapement has the movement inscribed J. L. Calame Robert Chaux de Fonds. It has day of the month indication. The fusee is reversed and is fitted with five-turn Geneva stop work. A chain guard is fitted to protect the carriage if there is a breakage in the chain. The lever escapement is right-angle club tooth. The balance is bimetallic. (c. 1880)

Plate 66 – An American watch with lever escapement but with the scape wheel at right-angles to the plates worm driven. The idea behind this was a long running simpler watch with fewer parts! The worm can be seen through the star-shaped hole in the top plate.

The dial is enamel. Movement inscribed New York Standard Watch Co. No. 29490. However, such a high torque was required to run the watch because of the worm pinion that this watch was a failure, and only about 26,000 were ever made and few of them survive today. (1887)

Plate 67 – This watch although bearing the name Bennett London is in fact a Glashütte watch by Lange. It is absolutely typical, with lever escapement, slits in the balance rim for tensioning the quarter screws, and the small screw that turns into a slot in the barrel arbor square to retain the ratchet wheel. The movement is numbered 6386. The case is later. (c. 1880)

Plate 68 – A German watch made in Dresden by the famous firm of A. Lange & Sohne Glashütte. The watch is numbered 29574 and has a lever escapement. It is a half quarter repeater and a fly back chronograph. It has an enamel dial. (c. 1890)

Plate 69 – This American watch has a crystal top plate and balance cock. It has a lever escapement and was made by A. W. Waltham & Co. No. 27. A letter written to an American in 1938 says that these watches were made fifty years before, and that one could still be obtained from Walthams for ten dollars. These watches must have been very difficult to jewel. Watches with crystal plates are very rare. This movement is in a later case. (c. 1888)

Plate 70 – This English watch is by Frodsham. It is an eight-day movement and has two barrels. The escapement is English lever. The bimetallic balance has a helical spring with duo in uno. The dial is enamel. The movement is inscribed Chas. Frodsham 84 Strand London FD FMsz. No. 06398. Frodshams were at this address up to 1897. The letter code indicates that the watch is of the highest quality. Much of Frodsham's work was made by the firm of Nicole Nielsen – whom they eventually bought in the 1930s. Nicole Nielsen supplied the finest watches of the day – indeed perhaps of any day. (c. 1890)

Plate 71 – This French cylinder watch is anonymous. The bars of the movement are so arranged as to spell out the word Paris. The dial is enamel. (Possibly 1890)

Plate 72 – This Roskopf watch is later than the previous example and has largely acquired the form that it was to keep for decades.

Plate 73 – This very flat Swiss watch has a right-angle club-tooth lever escapement. The bottom plate, the bridges, the barrel and the cocks are engraved all over. (Late 19th century)

Plate 74 – This Swiss tourbillon made by Fureur, and numbered 34394, is an inexpensive tourbillon. The centre of the back of the silver case is glazed to show the carriage. The balance is not of course at the centre of the carriage but this is of no importance. They are not of high quality and history does not relate how well they went, but without adjustments for isochronism the errors between the horizontal and vertical positions and those due to the watch running down would still be present. (c. 1920)

Plate 75 – This English watch is a Karrusel and was made by Yeomans. It is No. 85222. Britten lists a Joseph Yeomans of Cockermouth who died in 1905 aged sixty-five. This Karrusel watch has a sweep centre seconds hand, and gold minute and hour hands. The dial is enamel. The rough movement was made by B. Bonniksen and bears his patent No. 21421, and is of three-quarter-plate construction. The fourth wheel is at the centre. The going barrel has Geneva stop work. The carriage is revolved by a large pinion on the third arbor which meshes with the teeth on the edge of the carriage. The escapement is the English ratchet-toothed lever. The balance is two-arm bimetallic with gold screws and quarter nuts. The balance spring has seventeen and a half turns, is closely pitched and is free sprung with overcoil. The balance makes 18,000 beats an hour. The watch is keyless. (Early 20th century)

Plate 76 – This American watch was made by the Elgin National Watch Co. in 1918. It is 18 size with a 24-size dial. This precision watch was made to the highest possible standard for the United States Government. They were installed as navigational timepieces on destroyers – conventional marine chronometers were apparently unable to stand the vibration. Three hundred were made and sold to the Government at $275 each. This would then have been about £15 – a very high price. It is free sprung, has up and down indicator showing a reserve of forty hours. The balance is a Guillaume, and the escapement has diamond endstones. The number of jewels is twenty-one. The plates and keyless wheels are most beautifully damascened. The watch is engraved Father Time Elgin Ill. USA No. Adjusted 5 positions Safety Barrel 21 Jewel No. 21869977. The wheels are of low carat gold. The dial is enamel. Hand set is accomplished by means of a lever that is disclosed on removing the bezel. The watch was housed in a chronometer box with gimbals.

Plate 77 – This small Swiss minute repeater could be cased as a wrist watch, although it would no doubt have been a fob watch when first made. It has a club-tooth lever escapement. (c. 1910)

Plate 78 – This modern Swiss wrist watch shows the day, date and month, and the phases of the moon. Made by Matthey Tissot. It is also a chronograph with minute and hour recorder. (c. 1950)

Plate 79 – This English watch was the first successful self-winding wrist watch. Designed by John Harwood it has a compensation balance and lever escapement. Automatic winding is by means of a weighted segment which pivots at the centre and oscillates between buffer stops. The hands are set by rotating the milled bezel. (c. 1930)

Plate 80 – This self-winding watch prototype was made jointly by the author and P. W. Amis for S. Smith and Sons Ltd. The reduction between the self-winding weight and the barrel ratchet is obtained partly by a lever system and partly by gears. (1957)

Plate 81 – This American watch is the Hamilton 500A, the world's first successful electric wrist watch. It has a monometallic balance with special alloy spring to self-compensate and to resist magnetic fields. Impulse is given when contacts are made by the balance, due to the interaction of the magnetic fields caused in the coil on the balance and due to the permanent magnets beneath the coil. The battery has been removed but normally lies beneath the brass strap. (1957)

Plate 82 – The American Bulova tuning-fork watch. This marks the transition between the electric balance controlled watches to the quartz watch. Other tuning fork watches have been made by the Swiss. (1963)

Plate 83 – The Dynatron. This Swiss watch was the last of the balance controlled electronic watches. It had no contacts. The coil that does the triggering and the impulsing can be seen beneath the balance. The battery has been removed but lies normally beneath the cover bearing the name Dynatron. (1967)

Plates 84a and 84b – The dial of this self-winding chronograph has date work. This watch is called the Chronomatic. In the back view we can see the chronograph work. This is mounted complete on a sub-plate and this can be removed merely by undoing the three blued screws. This is an astonishing achievement. Note the unusual position of the winding button, this being at 9 o'clock. This watch is the result of a design exercise undertaken jointly by Buren Watch Co, Hamilton Watch Co, Breitling, Dubons and Dupraz, and Hever Leonidas. The performance is such that it can be rated to what the Swiss call 'chronometer' standard. The balance is large for such a watch, being 10·60 mm diameter. (1970)

Plate 85 – A Swiss watch, earrings and ring all matching. Made by Patek Philippe in 18 carat yellow gold set with diamonds, onyx and malachite. (1977)

Plate 86 – As Plate 85, but set with lapis lazuli and diamonds.

Bibliography

Baillie, G. H., *Clocks and Watches: an historical bibliography* (to 1800) (1951)

Baillie, G. H., *Watches: their history, decoration and mechanism* (Reprinted 1978 NAG Press)

Bassermann, Jordan *The Book of Old Clocks and Watches* (1964)

Berner, G. A., *Dictionnaire professionel illustré de l'horlogerie* (La Chaux de Fonds, 1961)

Britten, F. J., *Old Clocks and Watches and their Makers* ed. G. H. Baillie, Courtenay Ilbert, and Cecil Clutton, 8th edn, rev. Cecil Clutton (1973), 3rd imp. (1977)

Bruton, Eric, *Clocks and Watches 1400–1900* (1967)

Chamberlain, Paul M., *It's about Time* (1964)

Chapuis, A. and E. Jacquet, *The History of the Self Winding Watch* (Neuchâtel, 1952; London, 1956)

Chapuis, A. and E. Jacquet, *Technique and History of the Swiss Watch* (Bale, 1945) rev. 1970

Chapuis, A. and E. Jacquet, *La montre automatique* (Neuchâtel, 1952)

Chapuis, A. and E. Jacquet, *La montre chinoise* (Neuchâtel, 1919)

Cipolla, Carlo M., *Clocks and Culture 1300–1700* (1967)

Clutton, C., and G. Daniels, *Watches* (1965) new edition in prep.

Cumhaill, P. W. (Philip Coole), *Investing in Clocks and Watches* (1967)

Cuss, T. P., *The Camerer Cuss Book of Antique Watches*, ed. Terence Cuss (Woodbridge, 1976)

Cuss, T. P., *The Story of Watches* (1952)

Daniels, George, *Art of Breguet* (1974)

Daniels, George, *English and American Watches* (1967)

De Carle, Donald, *Complicated Watches and their Repair*, reissue (1977)

De Carle, Donald, *Practical Watch Repairing* (1946)

De Carle, Donald, *The Watch and Clock Encyclopedia* (1950), repr. (1976)

Fell, R. A., *Some Notes on the Balance and Spring* (1965)
Fried, H. B. *The Electric Watch Repair Manual* (New York, 1965) new edition (1972)
Fried, H. B., *The Watch Escapement: the lever, the cylinder, how to analyse, how to repair, how to adjust* (New York, 1974)
Gazeley, W. J., *Clock and Watch Escapements* ((1956), 3rd imp. (1975)
Gazeley, W. J., *Watch and Clock Making and Repairing* (1953), repr. (1975)
Glasgow, David, *Watch and Clock Making* (1885)
Gould, R. T., *The Marine Chronometer: its history and development* (1923)
Haswell, J. Eric, *Horology: the science of time measurement and the construction of clocks, watches and chronometers* (1928)
Howse, D., and B. Hutchinson, *Clocks and Watches of Captain James Cook 1769–1969* (1970)
Huber, Martin, *Die Uhren Von A. Lange and Sohne Glashütte/Sachen* (Munchen, 1977)
Jagger, Cedric, *Paul Philip Barraud* (1968)
Jaquet, E., and A. Chapuis, *Technique and History of the Swiss Watch from its beginnings to the present day* (Switzerland, 1953), facs. reissue (1970)
Laycock, W. S., *The Lost Science of John 'Longitude' Harrison* (Ashford, Kent)
Lecoultre, F., *A Guide to Complicated Watches* (Bienne, 1952)
Mercer, Vaudrey, *John Arnold and Son, chronometer makers* (Ramsgate, 1972–5)
Pellaton, James C., *Watch Escapements* (Switzerland, 1927), English Repr. (1949)
Pioneers of Precision Timekeeping, a symposium, Monograph No. 3, Antiquarian Horological Society (1965)
Rawlings, A. L., *The Science of Clocks and Watches* (New York, 1948)
Ullyett, Kenneth, *Watch Collecting for Amateurs* (1970)

Acknowledgements

The illustrations in this book are reproduced by courtesy of the following:

British Museum, London: Figs. 5, 6, 9, 10, 11, 12, 19 and Plates 1, 2, 3, 4, 5, 6, 7, 8, 9, 10, 11, 12, 13, 14, 15, 16, 17, 18, 19, 20, 21, 22, 23, 24, 25, 27, 28, 29, 30, 34, 35, 36, 37, 38, 39, 40, 41, 43, 45, 46, 47, 48, 50, 51, 52, 54, 55, 56, 57, 58, 59, 60, 61, 62, 63, 65, 66, 67, 68, 69, 70, 71, 73, 74, 75, 77

Graus Antiques: Plates 31, 32, 33, 42, 44, 53

Harold Malies: Fig. 28

National Maritime Museum, London: Plate 26

Patek Philippe S.A.: Plates 85, 86

SOME MUSEUMS WITH WATCH COLLECTIONS

ENGLAND

Bury St Edmunds Gershom-Parkington Memorial Collection

Greenwich National Maritime Museum

Liverpool City Museum

London British Museum

Science Museum

Victoria and Albert Museum

Wallace Collection

Museum of London (London Wall EC2)

Oxford Ashmolean Museum

Museum of History of Science

FRANCE

Besançon Musée d'Histoire des Beaux Arts

Paris Conservatoire des Arts et Métiers

Musée des Arts Decoratifs

Musée National du Louvre

Musée du Petit Palais

GERMANY

Munich Deutsches Museum

Stuttgart Württembegisches Landesmuseum

Dresden Staatliche Kimstsammlung 'Grunes Gewölbe'

HOLLAND

Amsterdam Rijksmuseum

ITALY

Milan Museo Poldi Pezzoli

SWITZERLAND

Basel Historisches Museum, Kirchgarten

Geneva Musée d'Horlogerie

La Chaux de Fonds Musée International d'Horlogerie

Le Locle Musée d'Horlogerie Chateau des Monts

Neuchatel Musée d'Art et d'Histoire

U.S.A.

CONNECTICUT

Bristol American Watch and Clock Museum

ILLINOIS

Rockford Time Museum

NEW YORK

New York Metropolitan Museum of Art

WASHINGTON D.C.

Smithsonian Institute

WALES

Cardiff National Welsh Museum

Index

INDEX TO COLOUR ILLUSTRATIONS

Figures in bold refer to the colour plates